Human Leukocyte Antigens - Updates and Advances

Edited by Sevim Gönen

Published in London, United Kingdom

Human Leukocyte Antigens - Updates and Advances
http://dx.doi.org/10.5772/intechopen.1001538
Edited by Sevim Gönen

Contributors
Sevim Gönen, Eman Farid, Penn Muluhngwi, Gizem Tumer, Andrés Jaramillo, Katrin Hacke, Beyza Goncu, Ali Osman Gurol, Tirasak Pasharawipas

First published in London, United Kingdom, 2023 by IntechOpen
IntechOpen is the global imprint of INTECHOPEN LIMITED, registered in England and Wales, registration number: 11086078, 5 Princes Gate Court, London, SW7 2QJ, United Kingdom

British Library Cataloguing-in-Publication Data
A catalogue record for this book is available from the British Library

Additional hard and PDF copies can be obtained from orders@intechopen.com

Human Leukocyte Antigens - Updates and Advances
Edited by Sevim Gönen
p. cm.
Print ISBN 978-1-83769-750-2
Online ISBN 978-1-83769-751-9
eBook (PDF) ISBN 978-1-83769-749-6

Meet the editor

Sevim Gönen obtained a master's degree and doctorate from the Faculty of Medicine, Department of Medical Microbiology, Gazi University, Turkey. She works as an academician and researcher in health sciences. She has more than 100 publications and 950 citations to her credit. She is a director for the Tissue-Typing Laboratory of Gazi University. She is a member of the Ankara Microbiology Association, the European Immunogenetics Federation (EFI), and the Society of Transplantation Immunology and Genetics (TIGED). Dr. Sevim Gönen has been a member of the Scientific Board of Transplantation Immunology of the Ministry of Health of Turkey since 2017.

Contents

Preface

Transplantation immunogenetics has always been and continues to be a subject of interest. In terminal-stage organ insufficiency and in some hematologic diseases, organ, tissue, and bone marrow transplantation are some of the treatment options. Transplantation immunology is an important part of organ transplantation. Organ transplantation can be successful thanks to the immunological harmony between the patient and the person who will donate the organ. For this purpose, some tests are applied to the patient and the donor in the pre-transplant period. Tissue compatibility antigens constitute one of the basic steps of organ transplantation in those terms.

This book discusses progress and advancements in human leukocyte antigens (HLAs) in six chapters. It highlights important points about HLA and transplantation immunogenetics.

Chapter 1, serves as an Introductory chapter to this Open Access book.

Chapter 2, "Human Leukocyte Antigens and Immune Tolerance" by Eman Farid, examines the relationship between immune tolerance and HLA antigens. Studies of HLA antigens and autoimmune diseases are the basis of population genetics.

Chapter 3, "Human Leukocyte Antigens and Transplantation" by Penn Muluhngwi and Gizem Tumer, provides an overview of HLAs in transplantation immunology, including information on crossmatch and antibody tests.

Chapter 4, "The Human Leukocyte Antigen System: Nomenclature and DNA-Based Typing for Transplantation" by Andrés Jaramillo and Katrin Hacke, discusses molecular tests and classification in immunogenetics of transplantation.

Chapter 5, "Tissue-Specific Immunity for Transplantable Endocrine Glands in the Context of HLA Expression" by Beyza Goncu and Ali Osman Gurol, emphasizes that HLA antigens are important not only in kidney and bone marrow transplants, but also in other organs.

Chapter 6, "Perspectives Concerning the Crucial Roles of MHC Molecules Corresponding to Individual Immunity against Emergent Viral Epidemics" was written by Tirasak Pasharawipas and shows that the MHC structure can be decisive in the body's response to infectious diseases, especially viral diseases.

I wish to thank the team at IntechOpen, especially Josip Knapic, for his support throughout the publication process.

Sevim Gönen
Associate Professor,
Gazi University,
Ankara, Turkey

Chapter 1

Introductory Chapter: Human Leukocyte Antigens – Updates and Advances

Sevim Gönen

1. Introduction

Organ transplantation has always been and continues to be a matter of curiosity. Transplantation of a healthy organ from a living donor or cadaver instead of an organ that cannot function in the body is called organ transplantation. The knowledge on immunology and immunosuppression has brought organ transplants to the agenda in the contemporary sense [1]. The most important aim of organ transplantation is to save the life of a person who is about to die due to organ failure, to increase the quality of life and to prolong its duration. Organ transplantation is accepted as a valid and advanced treatment method applied in many chronic organ diseases today. One of the basic steps of organ transplantation is constituted by tissue compatibility antigens.

HLA, which is called "Tissue Compatibility Antigens" or "Transplantation Antigens" necessary for the immune system to recognize foreign antigens, came from the abbreviation "Human Leukocyte Antigens" because they were first shown in leukocytes. The gene region encoding these human leukocyte antigens, which is necessary for the immune system to recognize self and non-self, is located on the sixth chromosome in humans and is called the Major Histocompatibility Complex Gene Region (MHC). Tissue compatibility antigens are also called MHC antigens. The discovery of the major tissue compatibility complex dates back to the 1940s [2]. The MHC consists of about 100 separate genes, and in this region, which is about 4000 kilobases in size, and there are some genes that have not been identified that are related to the immune response. MHC is divided into subregions as Class I, II, and III according to the characteristics of the encoded proteins [3–5].

HLA antigens encoded by polymorphic MHC genes are one of the most important elements of the immune system. Thanks to these, a bridge can be established between the natural immune system and the acquired immune system. In addition, the effective functioning of other elements of the acquired immune system, especially T cells, is provided by MHC molecules. From an evolutionary point of view, the sustained life on earth is, in a way, guaranteed thanks to the increased diversity of MHC alleles. However, such diversity in HLA alleles continues to be the most important factor in creating susceptibility or resistance to autoimmune and infectious diseases and preventing the success of organ transplantations.

HLA is an alloantigen that differs individually within the same species. In the following years of the discoveries, it was understood that they were found in all somatic cells, and it was shown that they have a very important role in tissue transplantation and that they determine the success of transplantation [6]. It was determined in the 1970s that people with a certain tissue type are more prone to certain diseases. In recent years, with the usage of recombinant DNA methods in which MHC molecules are crystallized, the base sequence of MHC complex genes and the amino acid sequences of MHC antigens that these genes provide synthesis have been revealed [7].

Clinical HLA applications are used for, transplantation, disease studies, anthropological studies, and paternity testing [8]. There are many studies proving the importance of HLA antigen compatibility, especially in bone marrow transplants [9, 10]. After the observation that MHC differences trigger tissue rejection, a lot of effort was spent to define these antigen differences. Unrelated donor sources are an important alternative for allogenic bone marrow transplantation, which is a curative treatment alternative in hematopoietic system malignancies and diseases where there is no suitable relative donor. An advanced HLA tissue bank system is needed to find an HLA-matched unrelated donor. The higher the capacity of HLA tissue banks, the higher the chance of finding a suitable donor. Again, in solid organ transplantation, tissue rejection has been tried to be minimalized by matching the HLA antigens of the donor and the recipient. Generally, the more alleles are compatible, the better the graft survival [11]. Studies have shown that MHC is associated with many diseases, especially infectious and autoimmune-based diseases. Such studies are carried out to determine disease risks and protective factors and to plan treatments [12].

Advances in technology and molecular diagnostic methods have facilitated the identification of HLA alleles that are important in organ and tissue transplants. Advantages of molecular methods: they are unique and flexible, new reagents can be developed as new alleles are identified, there are options that can perform studies with the desired sensitivity, they do not require live cells in studies, they are not affected by disease or treatment status of the individual, and they are more suitable for automation compared to serological and cellular methods, simultaneously. It can be summarized as being able to study a large number of samples, showing all the diversity in HLA genes and recognizing alleles that cannot be identified serologically.

Importance of MHC molecules in tissue transplantation: Immune responses after transplantation occur by well-defined mechanisms. Assuming that the recipient has not been sensitized before, the first responses include natural and non-antigen specific ones. Although natural immunity is not specific, it occurs rapidly and includes cellular elements such as neutrophils, macrophages, dentritic cells, NK and molecular elements such as toll like receptor(TLR), complement system, chemokine, cytokine [13]. The inflammation that occurs after the first encounter allows the antigen-presenting cells to mature and migrate to the lymphoid organs. In secondary lymphoid organs, dentritic cells initiate the acquired phase of the immune response by activating naive T cells. After the antigen-presenting cells activate the T cells, the acquired immune response phase begins. There are two ways in which alloantigens are recognized by the recipient's T cell. Indirect recognition, the recipient's T lymphocytes are triggered by alloantigens expressed on the donor antigen-presenting cell. These antigen-presenting cells are called passenger leukocytes, and they migrate from the transplanted organ to the recipient's secondary lymph nodes. These alloantigens are composed of endogenous peptides that have formed a complex with the donor MHC [14]. If the peptide presented with MHC is not an intrinsic peptide, it is perceived as a 'foreign' by the alloreactive T cell and leads to T cell activation. In indirect recognition, alloantigens on the donor cells are captured, processed by the recipient's cells and presented to the recipient T lymphocytes in the slit of the native MHC molecule.

DOI: http://dx.doi.org/10.5772/intechopen.1001371

The recognition of alloantigen by T cells ends with T cell activation, proliferation, elimination of the antigen. MHC molecules also appear to be responsible for post-transplant rejections.

Importance of MHC molecules in cancer: MHC molecules have an important role in tumor cells defeating the immune system. Tumor cells begin to produce new proteins called tumor antigens. These proteins, which are considered foreign by the immune system, are captured by APC, processed and presented to CD8+ T cells by MHC I molecules. As a result, a cytotoxic reaction against tumor cells begins. However, although tumor antigens are newly produced foreign antigens, since they originate from self-antigens, they are similar in structure to them and cannot stimulate the immune response very well. This causes the tumor cell to escape from the immune system. On the other hand, it is seen that MHC expressions are suppressed in tumor cells over time. In this way, tumor antigens are hidden from the immune system. This again causes tumor cells to escape from the immune system [15].

We wanted to highlight the important points about HLA with the knowledge and light of our academics who wrote the chapters of this book. Our academics are written by our departments by examining important topics to better illuminate, clarify, and guide future generations. We present our book chapter to all readers interested in transplantation immunology.

2. Conclusion

In conclusion, HLA antigens encoded by polymorphic MHC genes are one of the most important elements of the immune system. In addition to the high polymorphism feature, the codominant inheritance of MHC genes and the continuation of mutations in these genes cause a high level of immunological diversity among individuals in the population. It emphasizes its evolutionary importance as well as its contribution to the sustained continuation of life on earth. However, such diversity in HLA alleles continues to be the most important factor in creating susceptibility or resistance to autoimmune and infectious diseases and preventing the success of tissue transplantations. For this reason, better knowledge of HLA gene regions, structure, and function of HLA molecules will shed light on disease susceptibility, resistance formation, transplantation, and studies in the medical world.

We think you can find updates on HLA and transplantation in this book. We think we provide information that supports academics. The advancement of technology and science will continue to expand our knowledge.

Conflict of interest

The author declares no conflict of interest.

Author details

Sevim Gönen
Gazi University, Ankara, Turkey

*Address all correspondence to: sevgonen@gmail.com

References

[1] Aydogdu S, Cakır M, Arikan C, Tumgor G, Yuksekkaya HA, Yılmaz F, et al. Liver transplantation for progressive familial intrahepativ cholestasis: Clinical and histopathological findings, outcome and impact on growth. Pediatric Transplantation. 2007;**11**(6):634-640

[2] Snell GD. Studies in histocompatibility. Science. 1981;**213**:172-178

[3] Roitt IM, Delves PJ. Roitt's Essential Immunology. 10th ed. New York: Blackwell Science; 2001

[4] Roth DB. T cells and MHC. In: Male D, Brostoff J, Roth DB, Roith I, editors. Immunology. 7th ed. New York: Elsevier; 2006. pp. 105-124

[5] Dorak MT. Major histocompatibility complex. HLA review. Genes and Immunity. 2002;**10**:1-3

[6] Chinen J, Buckley RH. Transplantation immunology: Solid organ and bone marrow. The Journal of Allergy and Clinical Immunology. 2010;**125**:324-335

[7] Margulies DH. The major histocompatibility complex. In: Fundamental Immunology. Philadelphia: Lippincott Raven Puplishers; 1999

[8] Ghodke Y, Joshi K, Chopra A, et al. HLA and disease. European Journal of Epidemiology. 2005;**20**:475-488

[9] Little M, Marsh SGE, Madrigal JA. Current methodologies of human leukocyte antigen typing utilized for bone marrow donor selection. A Current Opinion in Hematology. 1998;**5**:419-428

[10] Reisner Y, Hagin D, Martelli MF. Haploidentical hematopoietic transplantation: Current status and future perspectives. Blood. 2011;**118**(23):6006-6017

[11] Eng HS, Leffell MS. Histocompatibility testing after fifty years of transplantation. Journal of Immunological Methods. 2011;**30**:1-21

[12] Kandemir O, Polat G, Sayıcı T, et al. Evaluation of MHC class 2 alleles in chronic hepatitis B patients and inactive hepatitis B carriers. Turkiye Klinikleri Journal of Medicine Science. 2010;**30**:1317-1323

[13] Huang Y, Rabb H, Womer KL. Ischemia–reperfusion and immediate T cell responses. Cellular Immunology. 2007;**248**(1):4-11

[14] Sayegh MH, Turka LA. The role of T-cell costimulatory activation pathways in transplant rejection. New England Journal of Medicine. 1998;**338**(25):1813-1821

[15] Biragyn A, Longo DL. Neoplastic "black ops": cancer's subversive tactics in overcoming host defenses. Seminars in Cancer Biology. 2012;**22**:50-59

Chapter 2

Human Leukocyte Antigens and Immune Tolerance

Eman Farid

Abstract

Immune tolerance is a known immune cascade of events by which our immune system can regulate its function, avoiding unwanted immune response reactions to immune privilege sites in our body. The role of HLA-G in fetal–maternal immune tolerance to prevent the embryo from being rejected can be applied to the process of transplantation as well as other clinical applications. The gut is also an important site of immune tolerance, with the constant assault of food antigens and its billions of resident microbes. In transplantation, the level of expression of HLA-G in the graft tissues correlates with organ acceptance and controls the recipient's immune response. Furthermore, tumor immune escape is associated with both the expression of immune checkpoint molecules on peripheral immune cells and soluble forms of the human leukocyte antigen-G (HLA-G) in the blood, which is consequently discussed as a clinical biomarker for disease status and outcome of cancer patients. Future studies are needed to explore more immune tolerance pathways for HLA-G and to apply and use this in transplantation to prevent rejection and treat miscarriage cases and autoimmune diseases. In addition, therapies to block HLA-G in malignant diseases are exciting and need more clinical trials. This chapter addresses and reviews the published articles related to the advances in HLA G and immune tolerance.

Keywords: HLA G, immune tolerance, transplantation, cytokines, immune system

1. Introduction

Immune tolerance is a known immune cascade of events by which our immune system can regulate its function, avoiding unwanted immune response reactions to immune privilege sites in our body. This regulation takes various mechanisms both at the molecular as well as cellular level, involving signaling, cell-to-cell interaction release of cytokines, and chemokines. Two interesting articles, addressing this topic were published in 2020, one on the immunology of the pregnant female [1] while the other on neuro-immunology [2].

2. The gut as a site of immune tolerance

With the constant assault of food antigens and its billions of resident microbes, the gut is an important site of immune tolerance. The mechanism and pathway between the immune system and the gut microbiota are still not totally revealed; interesting two articles were recently published addressing novel findings in this aspect; an article published titled “Novel antigen-presenting cell imparts Treg-dependent tolerance to gut microbiota” by Akagbosu et al. [3], their studies reveal parallel pathways for the establishment of tolerance to self and foreign antigens in the thymus and periphery, respectively, marked by the involvement of shared cellular and transcriptional programs. Another study titled: “ILC3s select microbiota-specific regulatory T cells to establish tolerance in the gut” by Lyu et al. [4]; their results define a paradigm whereby ILC3s select for antigen-specific RORγt + T reg cells, and against T helper 17 cells, to establish immune tolerance to the microbiota and intestinal health. Moreover, a review article on the same aspect was recently published titled “Localization and movement of T regs in the gastrointestinal tract” by Harad et al. [5].

3. Pregnancy and semi-allogeneic fetus acceptance

Pregnancy is a metabolic and immune challenge for the mother who, in her womb, has to adapt a semi-allogeneic fetus, whose 50 percent of antigens are of paternal origin; thus, the fetus should be viewed as an allograft and should be rejected via T cell-mediated, MHC-restricted mechanisms. The fetus does not come into direct contact with maternal tissue, but the embryonic trophoblast forms the interface between the maternal and fetal compartments and, thus, is the site of fetal antigen presentation [6]. Specific immune protective mechanisms are involved in establishing the active multifactorial maternal–fetal tolerance to the semi-allogenic fetus, where HLA-G plays an important role [7].

4. HLA-G and regulatory T cells in pregnancy

It has been demonstrated that mHLA-G expressed on trophoblasts is one of the key factors in regulating cytokine balance by shifting the Th1/Th2 balance toward Th2 polarization, a favorable medium for maintaining pregnancy [8]. The effects of HLA-G are highly concentration-dependent, and HLA-G-producing cells are located in the placental bed. This implies that sHLA-G might reduce the ability of T cells to function effectively in the pregnant uterus but is less potent in the periphery or away from the uterus [9]. HLA-G induces regulatory T cells by differentiating naïve T cells into $CD4^+$ $CD25^+$ or by forming temporary $HLA\text{-}G^+$ suppressor cells. Moreover, “$Foxp3^-$ suppressor T cells” are produced that function in the presence of IL-10. Also, $CD4^+$ $CD25^+$ $Foxp3^+$ and Th3 $IL\text{-}10^+$ $TGF\text{-}\beta^+$ are induced from naïve T cells when HLA-G inhibits the maturation of DCs. Maternal APC (macrophages and dendritic cells) are scattered all over the human decidualized endometrium during all stages of pregnancy. Studies showed that dendritic cells (DC) change to DC-10 by s HLA-G and IL-10and thus, induce production of type I T_{regs} cells (Tr1) followed by suppression of cytotoxic T cell responses by secreting IL10 and TGF-β. This plays an important role in maintaining maternal tolerance [10].

5. HLA-G molecule

In 1987, Geraghty and their Colleagues were the first who discover the HLA-G molecule, an 8–10 amino acid peptide, located within MHC loci at human chromosome

6q21.3 [11] and was extensively documented as a major potential promoter of tolerance at the human maternal–fetal interface; the expression of HLA-G was first described in trophoblasts [4]. HLA-G has 8escribed alleles and 3151 base pairs in [12]. The HLA-G molecule has special characteristics which make it different from other HLA class I molecules; it has restricted tissue distribution and lower polymorphisms. The HLA-G mRNA, encodes seven HLA-G isoforms: HLA-G1, HLA-G2, HLA-G3, and HLA-G4 membrane-bound proteins, and HLA-G5, HLA-G6, and HLA-G7 soluble proteins. Moreover, HLA-G membrane-bound isoforms have a shortened cytoplasmic tail which delays antigens recycling [13].

6. HLA-G polymorphism

HLA –G polymorphism is reduced with only 9 different HLA-G protein variants encoded by 28 alleles, of which 23 correspond to substitutions in the coding sequence. Not only polymorphism in non-coding regions affects HLA-G gene expression but also HLA G genotypes. Haplotype UTR-1 is associated with higher s HLA-G levels while haplotype UTR-5 or UTR-7 are associated with less s HLA-G levels [14].

7. Normal and pathological tissue distribution of HLA-G

It was initially thought that HLA-G under normal status is not expressed except by the fetal tissue cytotrophoblast, at the fetal–maternal interface, and in transplanted patients [15]. Further studies showed that in non-pathological (physiological) conditions HLA-G is also expressed in HLA class I-positive tissues such as oocytes, embryos, amnion, adult thymic epithelial cells, cornea, and nail matrix, which are considered immunologically privileged sites and in cytokine activated monocytes. Furthermore, HLA-G is expressed in pathological conditions, like some tumors, such as melanoma, colon carcinoma, lung carcinoma, ovarian carcinoma, gastric carcinoma, endometrial carcinoma, renal cell carcinoma, mesothelioma, breast carcinoma, trophoblastic tumors and hematopoietic tumors (hematologic malignancies, such as acute myeloid leukemia, chronic lymphocytic leukemia, and lymphoma) and may represent an escape mechanism from anti-tumoral immune responses [16]. HLA-G may be induced in other cell types during pathological processes, which include inflammatory disorders (e.g., skin inflammations and muscle inflammation), viral infections, HIV infection, by non-rejected allografts, and autoimmune diseases such as multiple sclerosis [MS] [17, 18].

8. Receptors for HLA-G

Unlike classical MHC-I proteins responsible for antigen presentation, HLA-G does not appear to play a role in activating the immune response. Instead, HLA-G exerts its inhibitory function against NK cells, T lymphocytes, and antigen-presenting cells (APCs) through direct binding to the inhibitory receptors. Three different inhibitory HLA-G receptors have been identified. These include the immunoglobulin-like transcript 2 (ILT-2 (LILRB1/CD85j)), which has been detected in monocytes, macrophages, dendritic cells, B cells, as well as subsets of T cells and NK cells; the immunoglobulin-like transcript 4 (ILT-4 (LILRB2/CD85d)) expressed by APC, namely monocytes, macrophages and dendritic cells, and the killer cell immunoglobulin-like receptor 2DL4 (KIR2DL4) (CD158d) which is mainly expressed by NK cells. Their expression might be increased by HLA-G binding itself. In addition, HLA-G has been shown to ligate the CD8 co-receptor expressed by certain T cells and NK cells [13].

The ITL2 and − 4 receptors are now known as LILRB1 and LILRB2 (leukocyte inhibitory receptors) [19]. Both receptors have broad specificities and bind classical MHC-I molecules in addition to HLA-G, however binding to HLA-G with a higher affinity than that with which they bind classical MHC-I proteins, indicating that ILT (LILR)/HLA-G interactions play a major role in controlling NK cell, T cell, and APC activity. Both receptors also have a higher affinity for HLA-G dimers, which are linked by an intermolecular disulfide bond. However, the affinity is different and dependent on the form (monomer, dimer). Shiroishi et al. demonstrated that HLA-G dimers induce more efficient ILT2-mediated signaling than monomers. A soluble form of HLA-G5 could form a disulfide-linked dimer with the intermolecular Cys42-Cys42 disulfide bond. In addition, the membrane-bound form of HLA-G1 can also form a disulfide-linked dimer on the cell surface of the Jeg3 cell line, which endogenously expresses HLA-G [20]. An important difference between the ILT2 and − 4 receptors is that HLA-G must associate with β2M to bind to the former. Nonetheless, both possess inhibitory properties and modulate the immune response accordingly. The KIR2DL4 receptor, unlike ILT2 and − 4, binds exclusively to HLA-G and not to classical MHC-I molecules. It has been shown to possess both inhibitory and stimulatory properties. As a result of the nature of KIR2DL4, the immunosuppressive effects of HLA-G have mostly been described through mechanisms involving the ILT2 and − 4 receptors [19].

9. HLA-G and immune tolerance

HLA-G induces immune tolerance by different mechanisms, mainly by regulation of cytokine production, suppression of CTL and NK cell killing activity and viability, inhibition of proliferation and induction of a suppressive phenotype in T helper cells, and alteration of DC stimulatory capacity and maturation of this lineage. Moreover, APCs transfected with HLA-G can prevent the proliferation of CD4 + T cells and drives an immunosuppressive profile where the cells produce high levels of TGF-β1 [21]. During these processes, HLA-G enhances the expression of Th2 anti-inflammatory cytokines, including IL-4, IL-10, and IL-13, and decreases Th1 pro-inflammatory cytokines, including IL-2, TNF-a, and IFN-g [22].

10. HLA-G and NK cell interactions

Many studies were done on the immunological effect of HLA-G on NK cell functions [23]; their results showed that HLA-G-positive target cells are protected from NK cytolysis through interaction with killing inhibitory receptors (KIR2DL4 and ILT2) HLA-G1 membrane-bound or soluble HLA-G protein [24]. ILT-2 is expressed on both NK and T cells, while KIRs, belongs to the immunoglobulin superfamily (Ig-SF), p49 Ig-SF KIR, and CD94/NKG2A inhibitory receptor, which belongs to the C-type lectin superfamily. Those receptors interact with HLA-E, which consequently inhibits NK lysis.

11. HLA-G and T cell function

Impairment of CD4+ and CD8+ T cell function has been well documented. Direct evidence is illustrated by the fact that HLA-G1, when transfected into target cells, blocks cytotoxic responses of CD8+ T cells specific for antigens expressed by these target cells [25]. Furthermore, soluble HLA-G has also been shown to induce apoptosis in CD8+ T cells by interacting with CD8, leading to Fas ligand (FasL) upregulation, FasL secretion, Fas/FasL interaction, then apoptotic signaling [26]. Furthermore, in vitro studies have demonstrated that soluble HLA-G5 inhibits CD4+ and CD8+ T Cell proliferation following an allogeneic response induced by T cell receptor activation, by binding to ILT2 receptors and arresting cell cycle progression.

A small fraction of CD4+ and CD8+ cells from peripheral blood have been found to stably express HLA-G and show less proliferation to allogeneic stimuli compared to their HLA-G- counterparts. These CD4 + HLA-G+ and CD8 + HLA-G+ cells represent novel Treg cell subsets, different from the traditional CD4 + CD25 + Foxp3+ population, and are independently able to suppress lymphocytic proliferation of both CD4 + HLA-G- and CD8 + HLAG- populations. Resting and activated CD4 T cells can rapidly acquire HLA-G1 through membrane exchange with HLA-G+ APC. Following the acquisition of membrane-bound HLA-G, these effectors can inhibit allo-proliferative responses. While these cells differ from CD4 + CD25 + Foxp3+ Treg cells, they temporarily function as such through the HLA-G1 they acquire but do not constitutively express. Acquisition of HLA-G1 by CD4+ T cells might explain how a few HLA-G+ cells can protect against immune aggression toward HLA-G target cells in the local milieu. There is, thus, considerable evidence demonstrating that in addition to directly inhibiting effector CD4+ and CD8+ T cells, HLA-G can generate populations of Treg cells to suppress these effectors [27].

12. Inhibition of antigen-presenting cell function

Interactions between recombinant HLA-G complexes and the ILT4 receptor on human dendritic cells, in vitro, resulting in impaired dendritic cell maturation characterized by reduced cell surface expression of MHC-II and co-stimulatory molecules typically induced by the maturation stimulus. The HLA-G/ILT4 interaction has also been shown to reduce the ability of dendritic cells to induce allogeneic T cell proliferation [19].

Recombinant HLA-G complexes have been shown to impair dendritic cell maturation, induce T cell energy, diminish CD4+ and CD8+ T cells responses, and generate Treg cells in ILT4 transgenic mice compared to their non-transgenic counterparts. Thus, transgenic animal models have demonstrated how HLA-G can impair APC maturation and, consequently, diminish their ability to activate T cells [28].

13. HLA-G in solid organ transplantation

HLA-G expression was detected in different solid organs after transplantation. In heart, liver, and combined liver-kidney transplant patients, increased sHLA-G has been associated with decreased acute rejection episodes, decreased chronic rejection, and a better transplant outcome [25]. HLA-G polymorphisms are associated with rejection or acceptance. Several clinical studies showed that the expression of

HLA-G has a protective role, induction of immune tolerance, and subsequent graft acceptance in transplantation. Moreover, studies showed that donor and recipient genotypes could influence the local HLA-G expression in the transplanted organ, as well as the activity of the host immune system response [22].

An interesting study from Japan conducted on 40 kidney transplant patients studied HLA-G expression on proximal tubular epithelial cells, and they concluded that HLA-G expression might confer long-term renal preservation effects in renal transplanted allografts [29].

Interestingly, another study showed that plasma levels of sHLA-G significantly decreased during the first year after renal transplantation and that lower levels of sHLA-G were found in recipients with post-transplant diabetes mellitus or obesity carrying the HLA-G14bpins/ins or HLA-G + 3142G/G genotypes [30].

Moreover, looking at HLA G genotypes, a study reported that HLA-G 3'-UTR variants are promising genetic predisposition markers both in donors and recipients that may help to predict susceptibility to either viral infectious complication of BKV or allograft rejection in kidney transplant [31].

Moving to heart transplants, a study evaluating HLA-G polymorphisms and cell-mediated rejection development, conducted on 55 recipients, concluded that HLA-G + 3196 G allele was identified as a risk factor for cell-mediated rejection diagnosis; according to their study, HLA-G may have a role in therapeutic/diagnostic strategies against heart transplant rejection [32].

A recent observational study on heart transplants, including 59 patients, concluded that soluble HLA-G levels decreased over the first year after a heart transplant. Also, higher HLA-G expression was associated with a higher frequency of infections but not with the burden of acute rejection or the development of coronary allograft vasculopathy, neither with the long-term patient or graft survival [33].

Regarding lung transplant, a single-center study examined 11 HLA-G SNPs in 345 consecutive recipients and 297 donors of a first bilateral lung transplant; specific donor SNPs were associated with mortality risk after lung transplantation, while certain donor–recipient SNP pairings modulated chronic lung allograft dysfunction risk. Trans-bronchial biopsies demonstrated predominantly epithelial, and therefore presumably donor-derived, HLA-G expression in keeping with these observations. This study is the first to demonstrate the effect of donor HLA-G SNPs on lung transplantation outcomes [34].

14. HLA-G and tumors

Tumor immune escape is associated with both the expression of immune checkpoint molecules on peripheral immune cells and soluble forms of the human leukocyte antigen-G (HLA-G) in the blood, which is consequently discussed as a clinical biomarker for disease status and outcome of cancer patients.

Regarding HLAG and T cell mechanism, a study demonstrated that priming of PBMC with sHLA-G1 protein before 48 h activation resulted in enhanced frequencies of ILT-2 expressing CD8C T cells, and in upregulation of immune checkpoint molecules CTLA-4, PD- 1, TIM-3, and CD95 exclusively on ILT-2 positive CD8C T cells. In contrast, when PBMC were primed with EV (containing HLA-G1 or not), upregulation of CTLA-4, PD-1, TIM-3, and CD95 occurred exclusively on ILT-2 negative CD8C T cells. Taken together, their data suggest that priming with s HLA-G forms induces a pronounced immunosuppressive/exhausted phenotype and that priming with sHLA-G1 protein or extra vesicular vesicles (EV) derived from HLA-G1 positive or negative SUM149 cells affects CD8C T cells complementary by targeting either the ILT-2 positive or negative subpopulation, respectively, after T cell activation. They report that they provide the first evidence that immune modulation by soluble HLA-G might involve IC molecules toward an immune suppressor phenotype. They conclude HLA-G functional analysis needs to be thoroughly performed in cancer patients [35].

Furthermore, Kataoka et al. published an interesting article addressing the killer immunoglobulin-like receptor 2D L4 (KIR2DL4) in pregnancy and in cancer metastasis by regulating mast cells [36]. Moreover, they added that stimulation of KIR2DL4 by HLA-G may enhance both conditions and may be useful in suppressing allergic reactions mediated by mast cells.

Regarding laryngeal carcinoma, a study suggested that HLA-G alleles may participate in LSCC pathophysiology; and The −14/−14 and − 14/+14 alleles may affect the biological function of the expressed HLA-G protein. Thus, their presence may act as a genetic risk factor that may predispose them to LSCC pathogenesis [37].

15. Therapy trials for s HLA-G for immune tolerance

Moving to therapeutic developments for immune tolerance, Radi et al. addressed in their review article "[Opinion on Immune Tolerance Therapeutic Development"] the major challenges to developing tolerance-inducing pharmaceutical drugs, including the selection of appropriate disease models to establish efficacy, adequate, and acceptable in vitro and in vivo safety assessments, relevant biomarkers of human safety and efficacy, and finally, some regulatory guidelines to successfully develop immune tolerance therapeutics [38].

Indeed multi-centric clinical trials are needed for HLA-G application in therapy. One line is to have s HLA-G as a therapy for repeated miscarriage cases, in the transplantation field, and in some autoimmune diseases. On the other side, therapies to block or downregulate the HLA-G molecule are needed in cases of tumors, thus enabling the immune system to fight those tumors and get rid of the tumor cells [13].

16. HLA G ongoing research

HLAG yet from the clinical diagnostic feasibility measuring it in serum is more practical as a non-invasive tool. Indeed using sHLAG to monitor post-transplant immune tolerance would be an easy way to monitor the possibility of losing such tolerance, thus alerting the clinician of a rejection episode before it occurs. Many researchers are working in this line, hoping to use sHLAG measurement as a diagnostic and monitoring laboratory tool in different clinical conditions [39]. A recent article by [40] addressed the role of the HLA G gene and its expression in the genesis of recurrent miscarriage. Another publication from India, addressed the role of stem cell transplant outcomes [41]. Measurement of sHLA-G plasma levels might be a good marker of efficient implantation after IVF [13]. Moreover, an excellent review article recently published in Frontier in Immunology reviewed the role of HLA G in organ transplantation [22].

Indeed the future is for immunotherapy and making use of the naturally occurring immune tolerance phenomenon that is, by time, more and more understood, whether in the gastrointestinal tract with the un-harmful gut microbiota or with the fetus during pregnancy, hopefully, can be achieved with a transplanted organ or as a therapy for cases of repeated abortion, autoimmune diseases and even in malignancy. Measuring T reg and s HLAG may be monitoring laboratory markers soon for several clinical conditions. In my opinion, the need for the development of new tools to analyze in-depth the HLA-G tumor neo-expression patterns, opening the way for the generation of new monoclonal antibodies and cell-based immunotherapies is urgent as mentioned by Loustau, et al. in their review [42].

17. Conclusion

The role of HLA-G in fetal–maternal immune tolerance to prevent the embryo from being rejected can be applied to the process of transplantation. In transplantation, the level of expression of HLA-G in the graft tissues correlates with organ acceptance and controls the recipient's immune response.

Future studies are needed to explore more immune tolerance pathways for HLA-G and to apply and use this in transplantation to prevent rejection and treat miscarriage cases and autoimmune diseases. In addition, therapies to block HLA-G in malignant diseases are exciting and need more clinical trials.

Author details

Eman Farid
Department of Microbiology and Immunology, College of Medicine, Arabian Gulf University, Kingdom of Bahrain

*Address all correspondence to: emanfarid57@gmail.com

References

[1] Dutta S, Sengupta P, Haque N. Reproductive immunomodulatory functions of B cells in pregnancy. International Reviews of Immunology. 2020;**39**(2):53-66. DOI: 10.1080/08830185.2019.1674299

[2] Negi N, Das BK. Decoding intrathecal immunoglobulins and B cells in the CNS: Their synthesis, function, and regulation. International Reviews of Immunology. 2020;**39**(2):67-79. DOI: 10.1080/08830185.2019.1711073

[3] Akagbosu B, Tayyebi Z, Shibu G, etal. Novel antigen-presenting cell imparts Treg-dependent tolerance to gut microbiota. Nature. 2022;**610**:752-760

[4] Lyu M, Suzuki H, Kang L, et al. ILC3s select microbiota-specific regulatory T cells to establish tolerance in the gut. Nature. 2022;**610**(7933):744-751. DOI: 10.1038/s41586-022-05141-x. Epub 2022

[5] Harada Y, Miyamoto K, Chida A, et al. Localization and movement of Tregs in gastrointestinal tract. Inflammation and Regeneration. 2022;**42**:47

[6] Hunt JSMRH, Petroff MG. Immunobiology of human pregnancy. In: Knobil and Neill's Physiology of Reproduction, Third ed. Vol. 2. 2006. p. 2759

[7] Tersigni C, Meli F, Neri C, Lacoangeli A, Franco R, Lanzone A, et al. Role of human leukocyte antigens at the feto-maternal interface in normal and pathological pregnancy: An update. International Journal of Molecular Sciences. 2020;**21**(13):4756

[8] Kanai T et al. Soluble HLA-G influences the release of cytokines from allogeneic peripheral blood mononuclear cells in culture. Molecular Human Reproduction. 2001;**7**(2):195-200

[9] Hunt JS, Petroff MG, McIntire RH, Ober C. HLA-G and immune tolerance in pregnancy. The FASEB Journal. 2005;**19**(7):681-693. Epub 2005/04/29

[10] Warning JC, McCracken SA, Morris JM. A balancing act: Mechanisms by which the fetus avoids rejection by the maternal immune system. Reproduction. 2011;**141**(6):715-724

[11] Kamishikiryo J, Maenaka K. HLA-G molecule. Current Pharmaceutical Design. 2009;**15**(28):3318-3324

[12] Robinson J, Barker DJ, Georgiou X, Cooper MA, Flicek P. Marsh SGE.IPD-IMGT/HLA database. Nucleic Acids Research. 2022;**48**:48-55. DOI: 10.1093/nar/gkz950

[13] Carosella ED, Moreau P, Lemaoult J, Rouas-Freiss N. HLA-G: From biology to clinical benefits. Trends in Immunology. 2008;**29**(3):125-132

[14] Martelli-Palomino G, Pancotto JA, Muniz YC, Mendes CT, Castelli EC, Massaro JD, et al. Polymorphic sites at 3 untranslated region of the HLA-G gene are associated with differential HLA-G soluble levels in the Brazilian and French population. PLoS One. 2013;**8**(10):e71742. DOI: 10.1371/journal.pone.0071742

[15] Carosella ED, Moreau P, Le Maoult J, Le Discorde M, Dausset J, Rouas-Freiss N. HLA-G molecules: From maternal-fetal tolerance to tissue acceptance. Advances in Immunology. 2003;**81**:199-252

[16] Yan WH. HLA-G expression in hematologic malignancies. Expert Review of Hematology. 2010;**3**(1):67-80

[17] Shih IM. Application of human leukocyte antigen-G expression in the diagnosis of human cancer. Human Immunology. 2007;**68**(4):272-276

[18] Kroner A, Grimm A, Johannssen K, Maurer M, Wiendl H. The genetic influence of the non-classical MHC molecule HLA-G on multiple sclerosis. Human Immunology. 2007;**68**(5):422-425

[19] Hunt JS, Morales PJ, Pace JL, Fazleabas AT, Langat DK. A commentary on gestational programming and functions of HLA-G in pregnancy. Placenta. 2007;**28**(Suppl. A):S57-S63

[20] Wu J, Zhang W, Hernandez-Lopez P, Fabelo E, Parikh M, Mulloy LL, et al. Isoforms of human leukocyte antigen-G and their inhibitory receptors in human kidney allograft acceptance. Human Immunology. 2009;**70**(12):988-994

[21] Guleria I, Sayegh MH. Maternal acceptance of the fetus: True human tolerance. Journal of Immunology. 2007;**178**(6):3345-3351

[22] Liu S, Bos NA, Verschuuren EAM, van Baarleand D, Westra J. Biological characteristics of HLA-G and its role in solid organ transplantation. Frontiers in Immunology. 2022;**13**:90209

[23] Riteau B, Rouas-Freiss N, Menier C, Paul P, Dausset J, Carosella ED. HLA-G2, −G3, and -G4 isoforms expressed as non-mature cell surface glycoproteins inhibit NK and antigen-specific CTL cytolysis. Journal of Immunology. 2001;**166**(8):018-5026

[24] Le Maoult J, Zafaranloo K, Le Danff C, Carosella ED. HLA-G up-regulates ILT2, ILT3, ILT4, and KIR2DL4 in antigen presenting cells, NK cells, and T cells. The FASEB Journal. 2005;**19**(6):662-664

[25] Naji A, Le Rond S, Durrbach A, Krawice-Radanne I, Creput C, Daouya M, et al. CD3+CD4low and CD3+CD8low are induced by HLA-G: Novel human peripheral blood suppressor T-cell subsets involved in transplant acceptance. Blood. 2007;**110**(12):39363948

[26] Contini P, Ghio M, Poggi A, Filaci G, Indiveri F, Ferrone S, et al. Soluble HLA-A,-B,-C and -G molecules induce apoptosis in T and NK CD8+ cells and inhibit cytotoxic T cell activity through CD8 ligation. European Journal of Immunology. 2003;**33**(1):125-134

[27] Feger U, Tolosa E, Huang YH, Waschbisch A, Biedermann T, Melms A, et al. HLA-G expression defines a novel regulatory T-cell subset present in human peripheral blood and sites of inflammation. Blood. 2007;**110**(2):568-577

[28] Horuzsko A, Lenfant F, Munn DH, Mellor AL. Maturation of antigen-presenting cells is compromised in HLA-G transgenic mice. International Immunology. 2001;**13**(3):385-394

[29] Kumano S, Okushi Y, Fujimoto K, et al. Role and expression of non-classical human leukocyte antigen-G in renal transplanted allografts. Clinical and Experimental Nephrology. 2021;**25**(4):428-438

[30] Piancatelli D, Maccarone D, Colanardi A, et al. HLA-G14bp ins/del polymorphism and post-transplant weight gain in kidney transplantation: Potential implications beyond tolerance. BMC Nephrology. 2020;**21**:109

[31] Rohn H, Schwich E, Michita RT, et al. HLA-G 3′ untranslated region gene variants are promising prognostic factors for BK polyomavirus replication

and acute rejection after living-donor kidney transplant. Human Immunology. 2020;**81**(4):141-146. DOI: 10.1016/j.humimm.2019.09.011

[32] Adamson MB, Ribeiro R, Yu F, et al. Human leukocyte antigen-G donor-recipient matching of the 14-base pair polymorphism protects against cancer after heart transplant. The Journal of Heart and Lung Transplantation. 2020;**39**(7):686-694

[33] Grille-Cancela Z, Barge-Caballero E, Suárez-Fuentetaja N, et al. Soluble HLA-G levels in heart transplant recipients: Dynamics and correlation with clinical outcomes. Transplant Immunology. 2023;**76**:101771

[34] Lazarte J, Ma J, Martinu T, et al. Donor human leukocyte antigen-G single nucleotide polymorphisms are associated with post-lung transplant mortality. The European Respiratory Journal. 2019;**54**:1802126

[35] Scwitch, et al, Soluble HLA-G and HLA-G bearing extracellular vesicles affect ILT-2 positive and ILT-2 negative CD8 T cells complementary. Frontiers in Immunology. 2020;**11**:2046. DOI: 10.3389/fimmu.2020.02046

[36] Kataoka TR, Ueshima C, Hirata M. Killer immunoglobulin-like receptor 2DL4 (CD158d) regulates human mast cells both positively and negatively: Possible roles in pregnancy and cancer metastasis. International Journal of Molecular Sciences. 2020;**21**:954. DOI: 10.3390/ijms21030954

[37] Barakat G, Elsharkawyb A, Nabiel A. Human leucocyte antigen – G gene polymorphism in laryngeal squamous cell carcinoma patients in Mansoura University hospitals. Egyptian Journal of Basic and Applied Sciences. 2021;**8**(1):214-221

[38] Radi ZA, Wynn TA. Opinion on immune tolerance therapeutic development. Toxicologic Pathology. 2020;**48**(6):712-717. DOI: 10.1177/0192623320948805

[39] Pistoia V, Morandi F, Wang X, Ferrone S. Soluble HLA-G: Are they clinically relevant? Seminars in Cancer Biology. 2007;**17**(6):469-479

[40] Bakleycheva MO, Bespalova ON, Ivashchenko T. The role of the HLA-G gene and its expression in the genesis of recurrent miscarriage. Journal of Obstetrics and Women's Diseases. 2022;**71**(1):101-108

[41] Kanga U, Mourya M, Seth T, Coshic P, Mahapatra M. Immunomodulatory role of HLA- G in hematopoietic stem cell transplantation outcome: A novel biomarker for GvHD. Transplantation. 2017;**101**(5S-3):S24. DOI: 10.1097/01.tp.0000520333.92669.64

[42] Loustau M, Anna F, Dréan R, et al. HLA-G Neo-expression on tumors. Frontiers in Immunology. 2020;**11**:1685

Chapter 3

Human Leukocyte Antigens and Transplantation

Penn Muluhngwi and Gizem Tumer

Abstract

The HLA (human leukocyte antigens) complex is positioned on chromosome 6 (6p21.3). The HLA genes respect the principles of Mendelian genetics and are co-dominantly expressed. The classical HLA genes are considered important for transplantation. HLA-A, -B, and -C are the classical HLA class I genes and are expressed in most of the somatic cells. HLA-DR, -DQ, -DP are the classical HLA Class II genes and are mainly expressed in antigen presenting cells such as B- cells, Activated T cells, Macrophages, Dendritic cells, and Thymic epithelial cells. In the presence of interferon, class II expression can be seen in other types of cells. This chapter includes the review of the structure and the function of the HLA molecule, and the most current HLA nomenclature. Subsequently evolution of HLA testing methodologies and advanced terminologies and techniques of HLA antibody evaluation that enhanced the bone marrow and solid organ transplantation is also discussed.

Keywords: HLA, nomenclature, HLA testing, transplantation, antibodies

1. Introduction

Class I and Class II molecules are structurally different. The basic structure of the Class I molecule is comprised of one polypeptide chain and a β2 microglobulin chain (encoded by a gene on chromosome 15). While, the HLA class II molecule has 2 polypeptide chains. Alpha1 and alpha2 domains form the peptide binding domain for class I whereas alpha1 and beta1 domains form the peptide binding domain for class II. The peptide binding domains contain the majority of polymorphic regions of the HLA antigen [1–3]. An essential function of the HLA class I molecule is to present peptides that are the products of the degradation of cytosolic proteins to the cell surface where they can be identified by the CD8 + T cells. CD8 receptor binds to the alpha3 region of the class I molecule and presents peptides to CD8 + T cells. HLA class II molecules present peptides that are the products of the degradation of *endocytosed* proteins to the cell surface where they can be identified by the CD4+ T cells. CD4 binds to the beta2 region of the class II molecule [1–3].

The rejection of transplanted tissue likely begins with the immune system recognizing differences in HLA antigens. The HLA system is considered to be second in importance to the ABO antigens in determining the success of solid organ transplants, but it is considered most important in the case of hematopoietic stem cell transplants [1, 3, 4].

2. Nomenclature

HLA nomenclature was updated in 2011 as the detection methods of HLA antigens dramatically changed since the discovery of the HLA system [1, 5, 6]. Originally, the polymorphisms of the Class I and Class II loci were solely based on serologic methods. With the rapid development and availability of molecular HLA typing methodologies, all HLA antigens are now uniformly named starting with the locus, antigenic specificity, and molecularly typed allele group. The asterisk "*" sign indicates that typing is performed by a molecular method and the colon ":" is a field separator.

Table 1 refers to molecular fields of HLA A*03:01:01:02 as an example and what they represent.

- A*03 is low resolution typing by molecular method which represents the first field digits by DNA-based nomenclature. In most cases this refers to the allele family or the serologic equivalent to the antigen such as serologic A3.

- A*03:01 is the allele and the 1st field (A*03) refers to a group of alleles that encode for the A3 antigen. 2nd field (:01) refers to an allele, which encodes a unique HLA protein (A*03:01).

- Intermediate-resolution typing includes a subset of alleles with the same first field antigen name and multiple possible different alleles and for which some alleles are excluded (ambiguous results). An example of an intermediate-resolution typing is A*03:01/20/26/37 which means the HLA type could be typed as A*03:01 or A*03:20 or A*03:26 or A*03:37 but not the other A*03 alleles.

- A group of alleles that specifies and encodes the same protein sequence for the peptide-binding region of an HLA molecule with the exclusion of non-expressed alleles on the cell-surface refers to high-resolution typing by molecular methodology. An example is A*03:26.

- 3rd field refers to a synonymous (silent) mutation, which represents a change in the DNA sequence without a change in the encoded protein. E.g. the difference between A*03:01:01 and A*03:01:02 is a synonymous (silent) mutation.

- The 4th field represents non-coding regions and after the 4th field are the expression modifiers (N, L, S, Q). Example includes Null alleles, which are alleles not expressed and denoted as capital N, L denotes low expression alleles, S is for secretory alleles, and questionable alleles are marked as Q.

Species	Locus	Antigen Equivalent	Allele	Silent Mutation	Outside exon	Expression
HLA	A*	03:	01:	01:	02	N, L, S, Q
	Locus*	1st Field	2nd Field	3rd Field	4th Field	

Table 1.
*HLA nomenclature: HLA A*03:01:01:02 as an example and what each field represent.*

Additionally, not all HLA typing technologies used today always allow for an unambiguous assignment of a single HLA allele. For these instances, strategies of common reporting of ambiguous strings can be used. Adding P and G codes provides easy reporting of ambiguous alleles. "P" code is added after the 2nd field for HLA alleles containing nucleotide sequences that encode the same protein sequence for the peptide binding domains (PBD). PBD for HLA class I alleles are exons 2 and 3, for HLA class II alleles exon 2 only. https://hla.alleles.org/alleles/p_groups.html "G" code is added after the 3rd field for HLA alleles that have identical nucleotide sequences across the exons encoding the PBD [7]. https://hla.alleles.org/alleles/g_groups.html

3. HLA typing

HLA typing is the ability to determine the HLA antigens of an individual prior to transplantation. The methodologies utilized for HLA typing have varied over the years from serological or cellular typing methods (that provided low resolution or antigen group of each locus) to the molecular methods (that utilize DNA sequence variation to provide intermediate to high resolution typing on a patient). Described below some of the HLA typing technologies previously (e.g. Complement-dependent cytotoxic assay) and currently used (including sequence specific priming; SSP, reverse Sequence specific oligonucleotide probe hybridization; rSSO, and sequence-based typing; SBT).

3.1 Complement-dependent cytotoxic (CDC) assay

The CDC assay (also known as the lymphocytotoxic or serological assay) was among the first assays used to determine HLA antigens (or type) of an individual. In this assay, individual's lymphocytes are incubated against a well-characterized panel of HLA antisera obtained from individuals sensitized to foreign HLA molecules by pregnancy or previous transplant and upon addition of rabbit serum, serving as a source of complement, cell injury is recorded by the microscopically characterizing cell membranes that take up vital or supravital dye [2, 8]. When more than 25–50% of cells are injured by a given antiserum (antibody), that population of cells is considered to have complementary antigens. An advantage of the technique is that it can be used to discern null alleles that are identified by molecular typing. Technical limitations that limit the results include the presence of unknown/rare alleles, difficulty in finding antisera to certain antigens or particular locus, homozygosity at one or more loci, and the assays inability to distinguish alleles that can only be characterized by DNA-based methods (such as HLA-DP alleles) [8, 9]. Further, CDC assays require optimization of the assay protocol including understanding the reactivity of each antisera, acceptable room temperature (22–25°C) conditions, evaluation of complement lots, avoiding contamination during pipetting and thorough mixing of assay components [2, 8, 9].

Variations or modifications of the CDC assay were later instituted to improve on its sensitivity. These modifications mostly occurred after the incubation step of anti-sera with cells such as addition of more wash steps (Amos-modified technique), addition of antihuman globulin (AHG).

3.2 DNA-based typing methods

DNA-based typing characterizes HLA polymorphism based on DNA sequence variation. Overcoming the initial restrictions of cost and incomplete primer and probes, DNA-based typing is available for every HLA lab to offer it to the transplantation community. Here are some advantages of this technology in comparison to CDC-based HLA typing: in DNA-based typing, a relatively small amount of DNA is used to perform the assay allowing for easier evaluation pediatric populations. There is no restriction of using only viable lymphocytes, the assay has the ability of accommodating variable DNA sources such as buccal swaps, whole blood, dried blood spots, cells tissues, and hair follicles, and the expression of relevant HLA antigens on the cell surface is not a requirement as in serological typing. Reagents such as DNA polymerase, primers, and probes are reproducibly synthesized and sold as standardized reagents to meet the needs of laboratory demand as oppose to antiserum in serological assays that can be in limited supply. Further, HLA typing requirements for Hematopoietic stem cell transplants (HSCT) require a minimum allele level typing which cannot be performed by serological methods [2, 9, 10].

Because DNA-based typing methods require DNA, DNA isolation technique that consistently provides adequate DNA quality and quantity from sample sources is very important. Methods of DNA isolation include Column techniques (Anion exchange columns), Salting out (cesium chloride gradient centrifugation or phenol extraction), and the use of magnetic beads (positive selection) [11]. Assessment of DNA quantity and quality can be done by UV Spectrophotometry with "Pure" DNA having a A260/280 absorption ratio of 1.8. Ratios greater than or lower than 1.8 suggest an increased presence of RNA and protein in the isolated sample respectively. There are also a fluorochrome-based methods for quantifying DSA. This is based on the principle that the fluorochrome will selectively bind double stranded DNA can be measured with a fluorometer [12].

Multiple factors can affect DNA quality. Acid Citric Dextrose (ACD) and Ethylenediaminetetraacetic acid (EDTA) are recommended as anti-coagulants for DNA isolation from a whole blood sample. Lithium heparin tubes should be avoided as lithium inhibits PCR. For recipients of hematopoietic stem cells, using whole blood may not be the best source to obtain accurate patient HLA typing as the patient's malignancy may include mutations at HLA loci that can lead to inaccurate results. Buccal swabs are a more suitable alternative. Samples received from patients with significant numbers of circulating cancer cells *e.g.* when the patient is in a leukemic blast phase, Allelic dropout can be seen at the time of HLA typing. In these situations, DNA obtained from a buccal swap is recommended. Recent blood transfusions may interfere with HLA typing as well.

Polymerase chain reaction (PCR) allowed for the development of simple, rapid, and improved characterization of allelic diversity at the HLA loci. In fact, next generation sequencing incorporates PCR. Therefore, HLA typing can be more accurate and precise (with high reproducibility and reliability). Amplified nucleotide sequences of a locus can reveal where and how alleles differ allowing for the evaluation of specific polymorphic amino acid residues in peptide binding and presentation.

Multiple PCR-based typing methods have been developed and applied to clinical HLA typing. Generally, the typing methods are such that the design of primer pairs targets polymorphic sequence motifs at a locus. The amplified sequence can then be analyzed by a variety of approaches including hybridization to an oligonucleotide probe (reverse sequence specific oligonucleotide-rSSO), digestion with restriction enzymes, chain termination sequencing reactions (Sequence-Based Typing-SBT) or evaluation of the mobility pattern using gel electrophoresis (sequence-specific primer-SSP). Other approaches use the specificity of the PCR itself with the 3′ end of the primer targeting the polymorphic site [10].

3.2.1 Sequence specific oligonucleotide (SSO)

The first PCR-based approach for HLA typing utilized labeled sequence specific oligonucleotide probe to hybridize onto amplified PCR products from a sample immobilized on a nylon or nitrocellulose membrane (the dot blot method) [13]. The SSO probes only bind to complementary sequence in the amplified DNA and can distinguish single nucleotide differences. Alleles in the samples could be identified by comparing patterns of probe reactivity to a panel of probes specific for informative sequence motifs [10]. Probes were labeled with either P^{32} (phased out), enzymes (such as horseradish peroxidase; HRP), digoxigenin or biotin. Biotin labeled probes can be detected with streptavidin conjugated to HRP (or alkaline phosphatase; AP) and a chromogenic or chemiluminescent substrate. Note that in the SSO probe typing approach a single PCR reaction amplifies all alleles at a target locus and the amplification wherever possible is locus specific.

SSO can be cumbersome with an increase in the number of probes and multiple separate hybridizations. To circumvent this limitation, the reverse hybridization approach (reverse dot blot) was developed wherein biotinylated amplified labeled PCR products are interrogated against an array of probes immobilized onto a solid support. Currently the most used reverse SSO (rSSO) assay for testing clinical samples is the bead-based rSSOP method where primers are used to amplify HLA gene polymorphic regions on Exons 2/3/4 for Class I and Exons 2/3 for Class II. In the PCR step, the amplicons are labeled with biotin. The amplicons are then applied to a panel of probes immobilized onto different uniquely fluorescently coded beads (identifiable by the Luminex technology). Note that, as opposed to SSO, the probes used in rSSO are not labeled. Each bead is coated with a unique allele- or group-specific oligonucleotide probe. Amplicons annealed to complementary probes are detected via streptavidin-phycoerythrin (SAPE) chemistry. The HLA typing is then deduced from the pattern of probe reactivity. Multiple probes must be used to characterize alleles that are amplified at each locus [10, 14].

The large number of probes and the complex probe reactivity patterns make it necessary to use computer programs for genotype analysis, which are updated periodically to incorporate newly identified alleles. In principle, with sufficient primers and probe, PCR-SSO should be capable of distinguishing all alleles. However, ambiguities may arise from the sharing of sequences between HLA alleles, the inclusion of new alleles into the typing system and software or when a probe reactivity pattern is consistent with more than one genotype. New probes are often developed to detect the newly defined alleles. Furthermore, SSO may not be as rapid as SSP. To ensure accuracy, control such as previously typed DNA, a positive control for each locus amplified and a no template negative control (molecular biology grade water) are typically included during assay setup [14].

3.2.2 Real-time PCR sequence specific priming (rtPCR-SSP)

rtPCR-SSP is based on the specificity of the primer extension. Historically it has been variously called allele-specific amplification (ASA) and amplification refractory mutation system (ARMS). In rtPCR-SSP, sequence specific primers are designed to amplify polymorphic regions of a sequence motif starting from the 3'end of the template. Amplification of targeted polymorphic sequences can be detected as bands on a gel. If there is no amplification or no product detected, the sample is assumed to lack the targeted motif. In such a case, a positive control included with test samples is examined to demonstrate successful PCR setup and exclude the potential of PCR inhibition and absence of template control. The positive control can be an unrelated monomorphic sequence that upon amplification produces a fragment distinguishable from the targeted polymorphic sequence post-gel electrophoresis. Although relatively fast for small sample size, SSP requires several PCR reactions to generate a typing resolution equivalent to intermediate or high-resolution typing. Thus, not well suited for large-scale throughput analysis.

In contrast to endpoint analysis of the PCR amplicons on a gel, modern rtPCR methodologies quantify the DNA during the exponential phase of the process. Two main fluorescent dyes, SYBR green and Tag Man, are used to detect PCR products in real-time instruments. In the PCR step, following primer extension and polymerization, SYBR green will intercalate into double stranded DNA and emit a strong fluorescent signal. When SYBR green is unbound, it exhibits little fluorescence. TagMan is a probe-based method. Here, probes labeled with a fluorophore (TagMan) and a quencher at opposite ends anneal to complementary DNA strands after the PCR denaturation step. During the extension phase, Taq DNA polymerase's 5'exonuclease activity cleaves the hybridized probe, releasing the reporter from the quencher and producing fluorescence that can be detected. These methods are fast, have a short turn-around time, require no post-amplification processing, and provide intermediate to high resolution typing, making them the most commonly used method for HLA typing in solid organ transplantation [10, 15].

3.2.3 Sequence-based typing

3.2.3.1 Sanger sequencing

Sequence-based typing is used to determine the exact nucleotide sequence of a gene or a region of a gene. It is considered the gold standard for HLA allele identification. Ideally sequence-based typing should result in unambiguous "high-resolution" typing (two field resolution e.g. A*02:01), describes the actual protein expressed, distinguishes null alleles, and detects new alleles. Strategies have evolved over time from the Maxam-Gilbert or Sanger sequencing wherein HLA type was determined by manual reading of radioactive tagged amplicons separated on a slab gel. The technique was modified to the Sanger capillary method that replaces the radioactive tags with fluorescent tags and automated reading of the amplicons separated by capillary gel electrophoresis. This modification made sequencing a more robust method for HLA typing, allowing for increased throughput, and decreased cost of the test. The first step in Sanger sequencing requires amplification (by PCR) of the antigen recognition site (specifically exons 2 and 3 for Class 1 and exon 2 for class II genes) using fluorescently tagged dideoxynucleotides. The amplicons (which make up both alleles of a locus) are then sequenced and compared to an HLA reference sequence in the reference data base (https://www.ebi.ac.uk/ipd/imgt/hla/) to determine allele assignments. Infrequently, alternative genotypes are obtained (*i.e.* ambiguous typing) wherein it is impossible to determine the phase of two or more polymorphisms at a locus and thus the level of resolution may not meet testing requirements. To resolve this, group-specific or allele-specific can be used in the PCR step to separate alleles prior to sequencing.

3.2.3.2 Next generation sequencing (NGS)

NGS is a newer methodology for high resolution typing that sequences large pieces of DNA or entire genomes. The sequence information is read as it is synthesized during the DNA synthesis reaction. Contrary to Sanger sequencing wherein a single DNA fragment is sequenced at a time, in NGS, millions of DNA fragments are simultaneously sequenced per run hence it is known as massive parallel sequencing. There are several NGS platforms that use different chemistry and detection methods, such as monitoring pH changes or capturing fluorescent signals to identify bases (A, C, G, T) as they are added one at time to the growing nucleotide chain. The fluorescent signals correspond to fluorescent tags on each of the four bases [16]. NGS has undergone several technological advances to a 2nd generation and a 3rd generation NGS (that allows for single molecule sequencing) [17].

Routine benchwork for NGS HLA typing is cumbersome and takes 2–3 days depending on the chemistry and detection method [18]. The first step in NGS usually involves amplifying the HLA gene region by PCR using primers that flank this region. The generated amplicons range in size from a single exon to all exons. The choice of the amplicon size depends on the starting material (amplification of short DNA segments is more robust than longer segments) and the target gene size. E.g. the HLA-DRB1 gene is three times longer than an HLA class I gene and amplification of this 15 kb DRB1 gene is difficult to achieve. Hence, HLA class II amplicons usually contain only a portion of the whole gene. Amplicon are kept relatively uniform in lengths for even representation during sequencing. Amplicons are then quantified, balanced, and pooled for library construction. The first step in library preparation is fragmentation. Since the generated amplicon reads usually range in size from 1000 to 6000 base pairs, these are cut into shorter fragments (200–300 base pairs) in a process known as fragmentation. Fragmentation can either be enzymic (e.g. restriction endonucleases) or physical (e.g. acoustical shearing-sound waves). Following fragmentation, oligonucleotides (adapters) are attached to the 5′ and 3′ ends of each amplicon fragment. The adapters have sequences that allow the fragments to bind to a solid support, serve as annealing sites for PCR and sequencing primers, and identify the source of the DNA (e.g. a loci or patient identifier). Adapters are "generic" (i.e. not specific for any gene) and same adapters are used for every library created. After adapter ligation the DNA is cleaned up to select for larger fragment sizes (by gel- or bead-based methods). Quantification by fluorometric or quantitative PCR (qPCR) methods can be used to ascertain the concentration of the DNA products. Multiple libraries can be pooled together and sequenced simultaneously as multiplex sequencing. Individual "barcode" sequences are added to each DNA fragment so each read can be identified and "binned" together with reads carrying the same. The created library is then sequenced on a solid surface (flow cell or bead) in the instrument sample chamber using a combination of lasers and fluorescent dyes. Using specialized software, the resulting sequence data is then analyzed, typically by aligning sequences (reads) from the sequencing data to a reference genome and comparing the obtained sequence to known HLA alleles in a database [17, 18].

Because NGS is a high throughput complex technique, many quality control checks have to be maintained to ensure generation of accurate and reliable results. It is recommended for each run should have an internal, positive, and negative controls. Good record keeping of the parameters for each sequencing run should be documented to show the quality of the run. Some parameters that can be monitored include cluster density, quality score, depth of coverage, read depth, heterozygosity (or the presence of two different alleles at a particular locus), and maintaining external proficiency testing.

4. Applications of HLA typing in transplantation

High resolution typing is commonly used to type recipients and unrelated donors in hematopoietic stem cell (HSCT) and cord blood (CBT) transplantation to treat leukemia, lymphoma, and other serious diseases affecting the hematopoietic system. For related donor selection high resolution typing is not a requirement but sequencing may be used to identify the best matched relative when similar alleles/antigens segregate within a family and/or the parents are unavailable.

In HSCT donor selection, preference is given to finding HLA-identical siblings. If no related donor is identified then the search for unrelated donors with matches at HLA-A, B, C, DRB1 is recommended. In fact, matching at these loci has been shown to increase survival following transplantation [19]. Matching for HLA-DQB1 and HLA-DRB3, -DRB4, and -DRB5 alleles is also recommended but not required while for the DPB1 loci, matching is for alleles with low immunogenic potential (permissive alleles). If a patient is to receive a mismatched transplant and determined to be sensitized (having HLA antibodies), finding the best matched donor may require typing of other donor loci to exclude donors to whom donor specific antibodies are present [20, 21]. This is done in an attempt to reduce the risk of graft failure. The requirement on matching for CBT is less stringent: a > 4/6 matches at HLA-A, B, DRB1 is the traditional match requirement [20].

Other center specific criteria to guide donor selection stem cell transplantation include killer immunoglobulin-like receptor (KIR) matching [22], HLA-DQ heterodimer (DQα-DQβ) matching [23], and HLA-B leader peptide matching. These parameters have each been reported to impact graft outcome [24]. The B leader peptide is an exon 1 encoded nonamer which when presented by HLA-E molecules stabilizes cell surface expression and enhances binding to CD94/NKG2-A on NK cells. Polymorphisms in the B leader peptide impact stability to HLA-E and binding to CD94/NKG2-A [25, 26]. In unrelated donor selection with mismatched HLA-B alleles, leader matched donors is associated with lower risk of GVHD compared to leader mismatched recipient/donor pairs [24]. Functional sequence variation at position 21 of Exon 1 of the leader peptide is also of importance. While this position is invariant for HLA-A and HLA-C, it is dimorphic in HLA-B and can either code for threonine (T) or Methionine (M) at the second position of the leader peptide. HLA B-matched pairs with MM leader genotype have worse outcomes (mortality and acute GvHD) relative to HLA B-matched pairs with TT leader genotype [27]. KIR genotyping offers an additional immunogenetic criteria for assessing haploidentical donors [22]. KIR genotyping is used to provide an assessment of Natural Killer (NK) cell alloreactivity. KIRs are polymorphic receptors on NK cells that interact with MHC class I molecules. On Interaction, NK cells become competent and able to activate itself against abnormal cells. NK cells without this interaction remain inactive through inhibitory receptors, specifically self-tolerant to autologous healthy cells. In allogenic HSCT, donors NK cells become alloreactive when inhibitory KIR receptors do not recognize recipient MHC Class I antigens [28]. This missing-self alloreactivity triggers Graft vs. leukemia (GVL) without promoting GVHD [29]. Thus, NK alloreactive donors (mismatched donors) will be preferred to donors having matched KIR ligands as recipients. It was recently reported that certain HLA-DQ heterodimers have lower risk than others; specifically donor and recipients matched for DQA1*02/03/04/05 paired with a DQB1*02/03/04 have lower relapse risk than other HLA-DQ heterodimer matched or mismatched pairs [23]. This opens additional avenues to risk categorize donors.

In addition to HLA compatibility, other considerations for donor selection in HSCT include Donor age (younger donors preferred), CMV matching, Donor gender (male donors preferred), stem cell source (Bone marrow vs. peripheral Blood), ABO compatibility, donor center location and specific center practices [30].

In solid organ transplantation (SOT), high resolution typing is not commonly used. This is because the priority for donor selection is avoidance of rejection and avoidance of donor specific antibodies in sensitized patient. Nevertheless, high resolution typing in SOT may guide donor section by allowing for more accurate interpretation of virtual crossmatches, DSA identification and selection of lower risk donors to whom the recipient does not have alloantibody [31].

5. HLA antibody testing

Microlymphocytotoxicity or CDC assay can be utilized to test serum against selected target cells to show the presence or absence of HLA antibodies. The testing is routinely done in HLA crossmatching. The assay is performed by testing serum from a potential recipient against lymphocytes from a potential donor. An enhanced variation of the CDC assay by adding antiglobulin reagent to increase the sensitivity has been used. These assays do not detect the specificity of the HLA antibodies.

5.1 Single antigen bead assay by Luminex

Current gold standard assay for HLA antibody testing is the Luminex (specialized flowcytometry) technology where each 5.6-micron polystyrene beads or microparticles are internally dyed with a unique combination of red and infrared dye. Utilizing of different intensities of the two dyes creates the bead set of 100 reactions by its unique spectral signature when excited by a laser beam. This permits multiplexing of up to 100 reactions in a single tube. Finally, the beads are coated with one recombinant HLA antigen or a haplotype of purified HLA antigen(s) (solid-phase methodology). Once the patient serum is exposed to the bead set, in the presence of an HLA antibody, the HLA antibody binds to its corresponding antigen on the bead. A fluorescent-labeled antihuman globulin (AHG) detects the antibody which produces a positive reaction and is measured by the mean fluoresce intensity (MFI) value. Flow cytometry and Luminex methods are more sensitive than CDC testing and detect the antibodies in the IgG form that may or may not fix complement. These assays are very sensitive. Therefore, the clinical significance of antibody detection by these methods should be interpreted along with a flow cytometric crossmatch prior to transplantation [1, 32, 33].

5.1.1 Detection and identification of HLA antibodies

Solid phase method can be utilized as an enzyme immunoassay where HLA antigen is bound to a solid phase plate or flow cytometer or Luminex detects the HLA antigen on the bead. Currently, three types of Luminex systems exist for HLA antibody detection and identification. These include a pooled HLA antigen system, a phenotype panel system, and a single antigen system. The pooled antigen system is mainly used as a screening tool for the presence or absence of HLA antibody. This assay is more sensitive but less specific. Reflex testing to identify the specificity of the HLA antigen can be performed to detect either by using the phenotype bead panel or a single antigen bead (SAB) testing. These assays have increased specificity.

The manufacturing of the pooled antigen, phenotype panel, and single antigen bead panel are different. The pooled antigen and phenotype panel use affinity columns to purify HLA antigens to coat the beads whereas the single antigen beads are coated by recombinant HLA antigens. A variation of SAB assay is also available to identify the antibodies that bind complement. This assay detects antibodies that bind complement (C1q or C3d). Donor specific HLA antibodies that have the capability to bind complement are associated with antibody mediated rejection and graft loss [1, 32, 33].

5.1.2 The advantages and disadvantages of antibody detection by Luminex technology

The Luminex technology provides fast, sensitive, and specific way of HLA antibody identification. These features of this assay significantly improved the clinical practice in the post-solid organ transplant setting, where the formation of new HLA antibodies to the graft can be identified quickly to determine a rejection episode or reduced graft function is reported. The speed and the specificity also aid for the utilization of apheresis to remove the antibodies to improve graft function in the post-transplantation setting. The Luminex methodology not only detects the specificity of the HLA antibody but also determines the strength and/or the avidity of the antibody which is as MFI value. Knowing the MFI value of antibodies present in the serum allows for risk stratification to the management of highly sensitized patients and proper adjustment of immunosuppressive medications [1, 32, 33].

There are some disadvantages with the Luminex technology. The assay is expensive and there are gaps in the antibody repertoire within the bead panels. If a patient has antibody where it is not represented in the bead panel, the assay provides false negative results. Also, standard Luminex kits do not provide information for the complement fixation capacity of the antibody. Since the HLA antibodies coating the single antigen bead assay are recombinant, presence of denatured antigens on beads may result in false positive results due to the cryptic epitopes becoming accessible during the manufacturing process. Inhibitory factors in patient serum may prevent antibody binding to single antigen bead (SAB) causing some beads to appear false negative. Treating the patient's serum with and dithiothreitol (DTT), ethylenediaminetetraacetic acid (EDTA), Heat inactivation or performing serial dilutions can resolve the issue. The single antigen bead assay is a semi-quantitative assay generally indicating more fluorescence as more antibodies bind to the beads. Meaning that if all of beads are occupied with HLA antibodies, there is the possibility that an excessive amount of HLA antibodies in a patient's serum may not be accurately detected. To ensure accurate assessment of antibody strength, titration studies are recommended. Also, if a patient has antibodies that recognize an epitope that is shared across multiple antigens, then the antibody may appear weaker than what it actually is. These shared epitope patterns or cross-reactive groups can create a false perception of compatibility, particularly when the antibody strength is not fully represented. To provide a more accurate assessment of HLA antibody strength, crossmatches can be used to performed [1, 31–33].

5.2 Flow Crossmatch

Since 1969, when Patel and Terasaki first demonstrated in their landmark paper the significance of recipient antibodies against donor antigens in mediating early rejection or graft loss in kidney transplantation [34], HLA laboratories now regularly perform crossmatches between donor cells and recipient serum. In their paper, Paul and Terasaki described the CDC assay discussed in the typing and antibody sections. The CDC crossmatch (CXM) assay is in essence an in vitro "surrogate" of what is to be expected from an allograft transplanted into the patient. Isolated lymphocytes are expected to express the same surface proteins as on allograft endothelium. The CDC crossmatch (CXM) assay has now morphed into the Flow crossmatch (FXM) assay. The FXM assay is a cell-based complement-independent assay where viable donor lymphocytes (T and B cells) are incubated with recipient serum to detect the donor-specific antibodies. Instead of quantifying the level of cell death as in CXM (an indicator of incompatibility), fluorescence is measured from a DSA-dependent fluorochrome conjugated secondary anti-IgG antibody. Lymphocytes can be isolated from lymph nodes or whole blood (using magnetic beads or ficoll/percoll methods) and pronase treated to reduce non-specific interactions. T cells and B cells are distinguished with different fluorescently antibodies specific to B and T lymphocyte surface proteins and florescence in the assay detected with a flow cytometer. The assay is a semi-quantitative method that reports results in terms of median channel shifts from a baseline or in relation to a set of MESF (molecules of equivalent soluble fluorescence) beads, and is more sensitive and less subjective than the visual assessment of cell death used in CXM [35, 36].

The Flow crossmatch assay is limited by interlaboratory variability as the technique and threshold for positivity varies between labs. Interactions in the FXM assay are not only HLA-specific nor donor specific. Therapeutic antibodies, specifically humanized and chimeric antibodies, can cause false positive reactions via non-specific immunoglobulin binding to Fc receptors. Such false positive FXM results can lead to denial of an otherwise acceptable donor organ. It is therefore essential that the crossmatch test occurs alongside extensive characterization of the patient's HLA antibody [35, 37].

Some Therapeutic interferences and consequences on Flow crossmatch:

- Rituximab (anti-CD20): False positive B cell crossmatch.
- Daclizumab (anti-CD25): False positive T and B cell crossmatch.
- Alemtuzumab (anti-CD25): False positive T and B cell crossmatch.
- Antithymocyte globulin (multiple targets): may or may not cause false positive reactions.
- IVIG (pooled human Ig): False positive T and B cell crossmatch at high concentrations.

5.3 Luminex crossmatch

In the Luminex crossmatch, isolated donor lymphocytes are coated onto class I and Class II specific beads. Captured monoclonals bind donor HLA antigens onto specific beads. HLA antigen-captured beads are then incubated with recipient sera. The presence of donor specific antibodies is detected with a Luminex analyzer after the addition of PE conjugated anti-human IgG antibody. The advantage of the Luminex crossmatch is that there is no requirement of viable lymphocytes, isolated donor cells can be frozen for future use and up to 91 assays can be performed at a time. Unfortunately, the ability for the Luminex crossmatch to detect DSA to HLA-DQ and DP is questionable [38].

5.4 Virtual crossmatch (VXM)

A virtual crossmatch (VXM) is the use of a patient's HLA-antibody profile to access compatibility with HLA antigens of a donor prior to a physical crossmatch and/or transplantation. It is not a physical test. Transplant programs have become increasingly reliant on VXMs to improve cold ischemic time and potentially reduce delayed graft function without increase risk of rejection [39] and offer the ability to evaluate low titer antibodies. The accuracy of a VXM is determined by the information used to perform the VXM. Factors such the ability of detect HLA antibodies in patient serum, the resolution level of Donor HLA typing and resolution of ambiguities, and the frequency and timeliness of HLA antibody testing [31, 40]. Having a more recently tested serum sample when performing improves the predictability of a VXM.

6. Applications of HLA antibody testing and Crossmatching in the clinical setting of solid organ transplantation

The heterogeneity in MHC region affords the immune system the ability to target diverse non-self-antigens (and pathogens). Development of anti-HLA antibodies occurs by three main mechanisms (listed in decreasing order sensitizing potential); prior transplantation, pregnancy, and blood transfusion. Patients having anti-HLA antibodies are described as sensitized. The presence of prior anti-HLA antibodies can result in rapid rejection of a transplanted organ or tissues; especially if there are preexisting donor specific anti-HLA antibodies (DSA). Detection and avoidance of anti-HLA antibodies is critical for successful transplantation. Further, well matched donor/recipient pairs have improved graft survival and limit sensitization in patients who may need retransplantion in future [41].

For patients not sensitized at transplantation, there is still the risk of developing de novo DSA or T- cell mediated rejection after transplantation. The likelihood of these events increases with the degree of mismatched antigens (i.e. antigens present in the donor but not in the recipient). The acceptable degree of matching and HLA typing resolution necessary for transplantation depends on the type of transplant (solid organ vs. stem cell) [41].

6.1 Allorecognition in transplantation

Allorecognition develops when the host T cells identify non-self-antigens and elicit for allograft rejection. There are two defined pathways for allorecognition. The "direct pathway" is involved with the presentation of the intact (non-processed) donor MHC peptide to host T cells while the presentation of processed peptides of the donor MHC molecules to host antigen presenting cells is referred as "indirect pathway." This second pathway requires T cell activation with co-stimulatory signals in addition to antigen recognition. In the alloimmune response, T Helper 1 (Th1) cells play a role in acute and chronic cellular rejection in solid organ transplantation and humoral antibody mediated rejection (AMR) is driven by Th2 cells. Once helper T cells activate the cytotoxic T cells, acute and chronic cellular rejection can be triggered [42].

6.2 HLA antibody production and graft survival

Individuals may generate anti-HLA antibodies when exposed to foreign HLA molecules through sensitizing events. Both pre-formed HLA antibodies and de novo donor specific antibodies are known to increase risk of rejection and premature allograft failure [43].

There are four major clinical types of rejection seen in kidney allografts where HLA antibodies are implicated: 1) hyperacute, 2) humoral/ antibody-mediated, 3) acute cellular, and 4) chronic [44].

Hyperacute rejection happens within minutes up to 24 hours after transplantation. Pre-formed donor specific HLA antibodies initiate a coagulopathy by binding HLA antigens expressed on the endothelium of the glomeruli and the graft microvasculature. This form of rejection has been virtually eliminated by the advanced technology of accurate HLA typing, timely identification of HLA antibodies, and enhanced techniques of crossmatching prior to transplantation. The absence of preformed HLA antibodies decreases the risk of hyperacute rejection and early humoral rejections. Currently no effective treatment is available for hyperacute rejection and is not reversible.

Humoral/antibody-mediated rejection often takes place within one [1] to three [3] months after transplantation. The patients may present with a rapid rise in the serum creatinine with an initial good baseline renal function. Pathology consists of the infiltration of endothelialitis, thrombosis, and C4d deposition in the peritubular capillaries and glomeruli. Development of de novo donor specific anti-HLA antibodies, and possibly non-HLA antibodies play a role in the pathogenesis. Apheresis, IVIG, and appropriate immunosuppressant therapy have been successful in the treatment of this form of rejection.

Acute cellular rejection usually presents within the first year after transplantation and more frequently within the first half of the year. The initiation and severity of the rejection depends on the type of immunosuppression used and patient compliance as well as the degree of matching between the recipient and the donor. A gradual rise in the serum creatinine with high blood pressure without significant decrease in urine output is the typical clinical presentation. This form of rejection usually has a favorable response to therapy. Pathogenesis involves the direct, and subsequently indirect, pathway of stimulations. With the availability of novel potent immunosuppressant drugs, the frequency of acute cellular rejection has significantly decreased. Degree of compatibility has an impact on the frequency of this type rejection. HLA identical living donor transplants, and 6- antigen-matched (HLA-A, -B, -DRB1) deceased donor transplants have a decreased chance of cellular rejection.

Chronic rejection may present months to years after transplantation. It occurs as an insidious rise in serum creatinine, proteinuria, and hypertension. This form of rejection does not have a favorable response to increased immunosuppression therapy, and despite treatment renal function continues to decline. Ultimately leads to a transplant glomerulopathy.

7. Advanced topics in HLA antibody analysis

Anti-HLA antibodies bind to immunogenic motifs, known as B cell epitopes and these epitopes can be shared across different HLA molecules within the same loci or across multiple loci. Each HLA molecule has multiple epitopes that can bind anti-HLA antibodies at once if these epitopes are spatially separated on the molecule. Advanced molecular HLA typing, now allows for better analysis of serologically defined HLA typing. This approach ultimately led to the recognition of HLA epitopes and the potential use for matching in transplantation [45].

7.1 Paratope: Epitope interactions

Complementarity-determining regions (CDR) determine the specific reactivity and affinity of an antibody. Paratope is the binding site of an antibody. B cell epitope is located within the CDRs of the antibody and is composed of 15–22 amino acid residues. Functional epitope is centrally located within the epitope and is made up of 2–5 amino acid residues within a 3-Å radius where the specificity of binding with the antibody occurs. The presence of at least one non-self-amino acid residue with the functional epitope allows for eliciting an immune response [45].

7.2 Defining epitopes

HLA B cell epitopes have been defined with two main methods: [1] Terasaki's serological epitopes (TerEps) that are identified by analysis of cross-reactivity patterns of anti-HLA antibodies using antibody absorption/elution technique, and [2] Rene Duquesnoy's HLAMatchmaker which defines theoretical epitopes based on differences in HLA amino acid sequences and modeling of molecular structure [45].

7.3 HLAMatchmaker

Exposure to foreign HLA antigens triggers HLA antibody production. The HLA antibodies bind to polymorphic amino acid residues known as epitopes on the antigens rather than binding to the whole HLA molecule. Each HLA molecule possesses many sites or epitopes for an antibody to bind. These epitopes may be private for only one HLA antigen or they may be shared across more than one HLA antigen (public) [46].

HLAMatchmaker (http://www.epitopes.net/) is an algorithm that analyzes HLA alleles structurally, considering these antigens as strings of unique molecular conformations that can be recognized by HLA antibodies critical in transplantation. This algorithm is based on experimentally antibody-verified epitopes described by polymorphic amino acids referred to as eplets.

Epitope repertoire was initially described based on serological cross-reactivity between HLA antigens and antibody specificities against private or public binding sites. This algorithm allows for defining the structural basis of HLA epitopes by looking at three-dimensional molecular structures and amino acid sequence differences between HLA antigens. In this model, HLA epitopes are determined by the polymorphic amino acid residues on the surface of the HLA molecule [47].

Most current version of HLAMatchmaker considers that antigenic proteins have functional epitopes where the amino acid residues are separated from each other by about 3 Ångstroms and at least one of amino acid residue is non-self. The current definition of an "eplet" is the areas of polymorphic residues within a radius of 3.0–3.5 Ångstroms. An epitope is described by the complete antigen–antibody interface (15 Ångstroms) composed of amino acids necessary for specificity and those that affect only affinity but not specificity. Utilizing HLAMatchmaker allows recipient and donor pair evaluations of humoral alloimmune response at the epitope level rather than antigen level humoral alloimmune response [46, 47].

An eplet can serve as a biomarker. A recent study from 2019 demonstrates that quantifying the number of single molecule eplet mismatches between individual class II HLA molecules (HLA-DR and HLA-DQ) may represent a precise, reproducible prognostic biomarker that can be utilized to modify immunosuppression based on individual patient risk [48].

7.4 Clinical applications of epitopes

Epitope-based matching provides a powerful matching tool to predict clinical outcomes. This tool can be used for decision-making at the time of organ allocation and dose adjustments of immunosuppression. In one study a strong linear correlation has been observed between DSA formation and the number of class I triplet mismatches in rejected post-transplant kidney grafts and also postpartum females [49].

A Canadian group in their kidney graft cohort demonstrated the association between class II epitope mismatch and DSA formation. Additionally, they found that the HLA-DR and DQ loci epitope mismatches were an independent risk factor for DSA formation [50].

The same group further characterized the epitope mismatches by defining an optimal threshold for each locus. HLA-DR (10 mismatched epitopes) and HLA-DQ (17 mismatched epitopes) were associated with a lower risk of developing DSA. On a separate cohort they subsequently investigated the synergistic impact of HLA epitope mismatch and patient non-compliance on graft survival. They concluded that increased number of epitope mismatches and patients with poor compliance had significant impact on graft loss when compared to low number of epitope mismatches [51].

Author details

Penn Muluhngwi[1] and Gizem Tumer[2,3]*

1 University of Minnesota, Minneapolis, Minnesota, USA

2 Indiana University, Indianapolis, IN, USA

3 Indiana University Health, Indianapolis, IN, USA

*Address all correspondence to: gtumer@iu.edu

References

[1] Klein J, Sato A. The HLA system. First of two parts. The New England Journal of Medicine. 2000;**343**(10):702-709

[2] Rodey G. HLA beyond Tears: Introduction to Human Histocompatibility: De Novo; 2000.

[3] Abbas AK, Lichtman AH, Pillai S. Cellular and Molecular Immunology. Elsevier Health Sciences; 2017

[4] Beasley L. Technical manual, 20th edition. Claudia S. Cohn, Meghan Delaney, Susan T. Johnson, Louis M. Katz, editors. Transfusion. 2021;**61**(2):634

[5] Tait BD. The ever-expanding list of HLA alleles: Changing HLA nomenclature and its relevance to clinical transplantation. Transplantation Reviews (Orlando, Fla.). 2011;**25**(1):1-8

[6] Fung MK, Benson K. Using HLA typing to support patients with cancer. Cancer Control. 2015;**22**(1):79-86

[7] Marsh SG, Albert ED, Bodmer WF, Bontrop RE, Dupont B, Erlich HA, et al. Nomenclature for factors of the HLA system, 2010. Tissue Antigens. 2010;**75**(4):291-455

[8] Schreuder GM, Hurley CK, Marsh SG, Lau M, Fernandez-Vina MA, Noreen HJ, et al. HLA dictionary 2004: Summary of HLA-A, -B, -C, -DRB1/3/4/5, -DQB1 alleles and their association with serologically defined HLA-A, -B, -C, -DR, and -DQ antigens. Human Immunology. 2005;**66**(2):170-210

[9] Tinckam K. Histocompatibility methods. Transplantation Reviews (Orlando, Fla.). 2009;**23**(2):80-93

[10] Erlich HA, Opelz G, Hansen J. HLA DNA typing and transplantation. Immunity. 2001;**14**(4):347-356

[11] Ali N, Rampazzo RCP, Costa ADT, Krieger MA. Current nucleic acid extraction methods and their implications to point-of-care diagnostics. BioMed Research International. 2017;**2017**:9306564

[12] Ahn SJ, Costa J, Emanuel JR. PicoGreen quantitation of DNA: Effective evaluation of samples pre- or post-PCR. Nucleic Acids Research. 1996;**24**(13):2623-2625

[13] Saiki RK, Bugawan TL, Horn GT, Mullis KB, Erlich HA. Analysis of enzymatically amplified beta-globin and HLA-DQ alpha DNA with allele-specific oligonucleotide probes. Nature. 1986;**324**(6093):163-166

[14] Rajalingam R, Cecka M, Reed EF. Chapter 30 - molecular HLA typing methods used in clinical laboratories. In: Grody WW, Nakamura RM, Strom CM, Kiechle FL, editors. Molecular Diagnostics. San Diego: Academic Press; 2010. pp. 367-379

[15] Arya M, Shergill IS, Williamson M, Gommersall L, Arya N, Patel HR. Basic principles of real-time quantitative PCR. Expert Review of Molecular Diagnostics. 2005;**5**(2):209-219

[16] Erlich HA. HLA typing using next generation sequencing: An overview. Human Immunology. 2015;**76**(12):887-890

[17] Profaizer T, Kumánovics A. Human leukocyte antigen typing by next-generation sequencing. Clinics in Laboratory Medicine. 2018;**38**(4):565-578

[18] Smith AG, Pereira S, Jaramillo A, Stoll ST, Khan FM, Berka N, et al. Comparison of sequence-specific

oligonucleotide probe vs next generation sequencing for HLA-A, B, C, DRB1, DRB3/B4/B5, DQA1, DQB1, DPA1, and DPB1 typing: Toward single-pass high-resolution HLA typing in support of solid organ and hematopoietic cell transplant programs. HLA. 2019;**94**(3):296-306

[19] Lee SJ, Klein J, Haagenson M, Baxter-Lowe LA, Confer DL, Eapen M, et al. High-resolution donor-recipient HLA matching contributes to the success of unrelated donor marrow transplantation. Blood. 2007;**110**(13):4576-4583

[20] Dehn J, Spellman S, Hurley CK, Shaw BE, Barker JN, Burns LJ, et al. Selection of unrelated donors and cord blood units for hematopoietic cell transplantation: Guidelines from the NMDP/CIBMTR. Blood. 2019;**134**(12):924-934

[21] Ciurea SO, Cao K, Fernandez-Vina M, Kongtim P, Malki MA, Fuchs E, et al. The European Society for Blood and Marrow Transplantation (EBMT) consensus guidelines for the detection and treatment of donor-specific anti-HLA antibodies (DSA) in Haploidentical hematopoietic cell transplantation. Bone Marrow Transplantation. 2018;**53**(5):521-534

[22] Dhuyser A, Aarnink A, Pérès M, Jayaraman J, Nemat-Gorgani N, Rubio MT, et al. KIR in allogeneic hematopoietic stem cell transplantation: Need for a unified paradigm for donor selection. Frontiers in Immunology. 2022;**13**:821533

[23] Petersdorf EW, Bengtsson M, Horowitz M, McKallor C, Spellman SR, Spierings E, et al. HLA-DQ heterodimers in hematopoietic cell transplantation. Blood. 2022;**139**(20):3009-3017

[24] Petersdorf EW, Stevenson P, Bengtsson M, De Santis D, Dubois V, Gooley T, et al. HLA-B leader and survivorship after HLA-mismatched unrelated donor transplantation. Blood. 2020;**136**(3):362-369

[25] Braud V, Jones EY, McMichael A. The human major histocompatibility complex class Ib molecule HLA-E binds signal sequence-derived peptides with primary anchor residues at positions 2 and 9. European Journal of Immunology. 1997;**27**(5):1164-1169

[26] Braud VM, Allan DS, O'Callaghan CA, Söderström K, D'Andrea A, Ogg GS, et al. HLA-E binds to natural killer cell receptors CD94/NKG2A, B and C. Nature. 1998;**391**(6669):795-799

[27] Hallner A, Bernson E, Hussein BA, Ewald Sander F, Brune M, Aurelius J, et al. The HLA-B -21 dimorphism impacts on NK cell education and clinical outcome of immunotherapy in acute myeloid leukemia. Blood. 2019;**133**(13):1479-1488

[28] Moretta A, Bottino C, Vitale M, Pende D, Biassoni R, Mingari MC, et al. Receptors for HLA class-I molecules in human natural killer cells. Annual Review of Immunology. 1996;**14**:619-648

[29] Locatelli F, Pende D, Falco M, Della Chiesa M, Moretta A, Moretta L. NK cells mediate a crucial graft-versus-Leukemia effect in Haploidentical-HSCT to cure high-risk acute Leukemia. Trends in Immunology. 2018;**39**(7):577-590

[30] Ayuk F, Balduzzi A. Donor selection for adults and pediatrics. In: Carreras E, Dufour C, Mohty M, Kröger N, editors. The EBMT Handbook: Hematopoietic Stem Cell Transplantation and Cellular Therapies. Cham (CH): Springer; 2019. pp. 87-97

[31] Pinelli DF, Tambur AR. Virtual crossmatching for deceased donor

transplantation: One size does not fit all. Kidney International. 2020;**97**(4):659-662

[32] Roberts T, Tumer G, Gebel HM, Bray RA. Solid-phase assays for the detection of alloantibody against human leukocyte antigens: Panacea or Pandora? International Journal of Immunogenetics. 2014;**41**(5):362-369

[33] Tumer G, Gniadek T, Baye J, Pena R, Warner P, Fung M, et al. The effect of serum pretreatment regimens for the detection of HLA class I antibodies in platelet-refractory patients. Transfusion. 2020;**60**(3):488-497

[34] Patel R, Terasaki PI. Significance of the positive crossmatch test in kidney transplantation. The New England Journal of Medicine. 1969;**280**(14):735-739

[35] Jaramillo A, Ramon DS, Stoll ST. Technical aspects of Crossmatching in transplantation. Clinics in Laboratory Medicine. 2018;**38**(4):579-593

[36] Alheim M, Wennberg L, Wikström AC. Pronase independent flow cytometry crossmatching of rituximab treated patients. Human Immunology. 2018;**79**(2):132-135

[37] Desoutter J, Apithy MJ, Guillaume N. Unexpected positive prospective Crossmatches in organ transplant. Experimental and Clinical Transplantation. 2017;**15**(3):253-259

[38] Billen EVA, Voorter CEM, Christiaans MHL, Van Den Berg-Loonen EM. Luminex donor-specific crossmatches. Tissue Antigens. 2008;**71**(6):507-513

[39] Rohan VS, Pilch N, Moussa O, Nadig SN, Dubay D, Baliga PK, et al. Virtual Crossmatching in kidney transplantation: The wait is over. Journal of the American College of Surgeons. 2020;**230**(4):373-379

[40] Tambur AR, Ramon DS, Kaufman DB, Friedewald J, Luo X, Ho B, et al. Perception versus reality?: Virtual crossmatch--how to overcome some of the technical and logistic limitations. American Journal of Transplantation. 2009;**9**(8):1886-1893

[41] Montgomery RA, Tatapudi VS, Leffell MS, Zachary AA. HLA in transplantation. Nature Reviews. Nephrology. 2018;**14**(9):558-570

[42] Game DS, Lechler RI. Pathways of allorecognition: Implications for transplantation tolerance. Transplant Immunology. 2002;**10**(2-3):101-108

[43] Molina J, Navas A, Agüera ML, Rodelo-Haad C, Alonso C, Rodríguez-Benot A, et al. Impact of preformed donor-specific anti-human leukocyte antigen antibody C1q-binding ability on kidney allograft outcome. Frontiers in Immunology. 2017;**8**:1310

[44] Braun WE. Renal transplantation: Basic concepts and evolution of therapy. Journal of Clinical Apheresis. 2003;**18**(3):141-152

[45] Sypek MP, Hughes P, Kausman JY. HLA epitope matching in pediatric renal transplantation. Pediatric Nephrology. 2017;**32**(10):1861-1869

[46] Duquesnoy RJ. Reflections on HLA epitope-based matching for transplantation. Frontiers in Immunology. 2016;**7**:469

[47] Lemieux W, Mohammad hassanzadeh H, Klement W, Daniel C, Sapir-Pichhadze R. Matchmaker, matchmaker make me a match: Opportunities and challenges in

optimizing compatibility of HLA eplets in transplantation. International Journal of Immunogenetics. 2021;**48**(2):135-144

[48] Wiebe C, Kosmoliaptsis V, Pochinco D, Gibson IW, Ho J, Birk PE, et al. HLA-DR/DQ molecular mismatch: A prognostic biomarker for primary alloimmunity. American Journal of Transplantation. 2019;**19**(6):1708-1719

[49] Dankers MK, Witvliet MD, Roelen DL, de Lange P, Korfage N, Persijn GG, et al. The number of amino acid triplet differences between patient and donor is predictive for the antibody reactivity against mismatched human leukocyte antigens. Transplantation. 2004;**77**(8):1236-1239

[50] Wiebe C, Pochinco D, Blydt-Hansen TD, Ho J, Birk PE, Karpinski M, et al. Class II HLA epitope matching-a strategy to minimize de novo donor-specific antibody development and improve outcomes. American Journal of Transplantation. 2013;**13**(12):3114-3122

[51] Wiebe C, Nevins TE, Robiner WN, Thomas W, Matas AJ, Nickerson PW. The synergistic effect of class II HLA epitope-mismatch and nonadherence on acute rejection and graft survival. American Journal of Transplantation. 2015;**15**(8):2197-2202

Chapter 4

The Human Leukocyte Antigen System: Nomenclature and DNA-Based Typing for Transplantation

Andrés Jaramillo and Katrin Hacke

Abstract

The major histocompatibility complex in humans, known as the human leukocyte antigen (HLA) and located on chromosome 6, is the most polymorphic genetic system in humans. The biological role of the HLA class I and class II molecules is to present processed peptides to CD8+ and CD4+ T lymphocytes, respectively. These cells can also respond to foreign (allogeneic) HLA molecules (direct allo-recognition) or to foreign HLA-derived peptides (indirect allo-recognition), respectively. Thus, the HLA system controls the acceptance or rejection of transplanted foreign tissues and organs (allografts). High-resolution HLA typing is routinely performed to provide HLA matching in hematopoietic stem cell transplantation to prevent allograft rejection and graft-versus-host disease. In contrast, low-resolution HLA typing is routinely performed in solid organ transplantation to provide HLA matching but, most importantly, to allow for the detection of donor-specific antibodies to prevent antibody-mediated allograft rejection. The capability to amplify DNA by polymerase chain reaction has facilitated the clinical application of molecular techniques and, currently, several molecular HLA typing methods are now available in the histocompatibility laboratory. Herein, we describe the different molecular HLA typing techniques and the different levels of HLA typing resolution used for clinical purposes.

Keywords: human leukocyte antigen, transplantation, nomenclature, genetics, antigen recognition

1. Introduction

The major histocompatibility complex (MHC) in humans, known as the human leukocyte antigen (HLA) system, controls the acceptance or the rejection of transplanted foreign tissues and organs (allografts). Immunocompetent CD8+ and CD4+ T lymphocytes respond to foreign (allogeneic) HLA molecules (direct recognition) or to foreign HLA-derived peptides (indirect recognition), respectively [1, 2]. Of note, allorecognition of HLA molecules may activate up to 10% of the total T lymphocyte pool. The extent of T lymphocyte activation varies according to the level of HLA disparity between donor and recipient: the greater the level of HLA disparity, the stronger the extent of the response.

HLA class I molecules (A, B, C) are expressed on all nucleated cells and are recognized by CD8+ T lymphocytes. Alternatively, HLA class II molecules (DR, DQ, DP) are only expressed on the surface of antigen-presenting cells and are recognized by CD4+ T lymphocytes (**Table 1**) [1, 2]. The HLA class II molecules are also expressed in a variety of other cell types, such as endothelial cells, upon stimulation with interferon-γ. Every individual inherits nine clinically relevant HLA alleles (three HLA class I and six HLA class II) from each parent. Such antigens are co-dominantly expressed on the cell surface. The entire set of clinically relevant HLA-A, B, C, DRB1, DRB3/B4/B5, DQA1, DQB1, DPA1, and DPB1 genes encoded on chromosome 6 is called a "haplotype" (**Figure 1**). According to Mendelian laws, the HLA "haplotypes" are inherited en block from each parent (**Figure 2**). This concept is of high importance in clinical hematopoietic stem cell transplantation (HSCT) where the first donor of choice is an HLA identical sibling. Based on the Mendelian laws of inheritance, everyone has only a 25% probability of being HLA identical to their siblings (**Table 2**).

Region	Gene products	Tissue location	Function
Class I	HLA-A, B, C	Nucleated cells	Recognition of tumor and virus-infected cells by CD8+ T lymphocytes
Class II	HLA-DR, DQ, DP	Antigen-presenting cells: B lymphocytes, macrophages, dendritic cells, endothelial cells	Recognition of foreign antigens by CD4+ T lymphocytes

Table 1.
Expression of HLA genes.

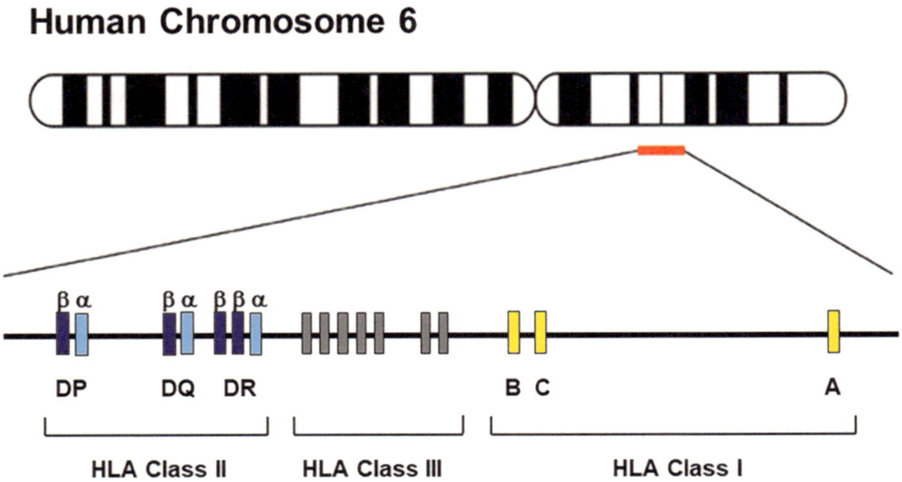

Figure 1.
The human leukocyte antigen complex on chromosome 6. The HLA class I region is 3-6 kb long and the HLA class II region is 4-11 kb long. The HLA class III is not part of the polymorphic HLA system.

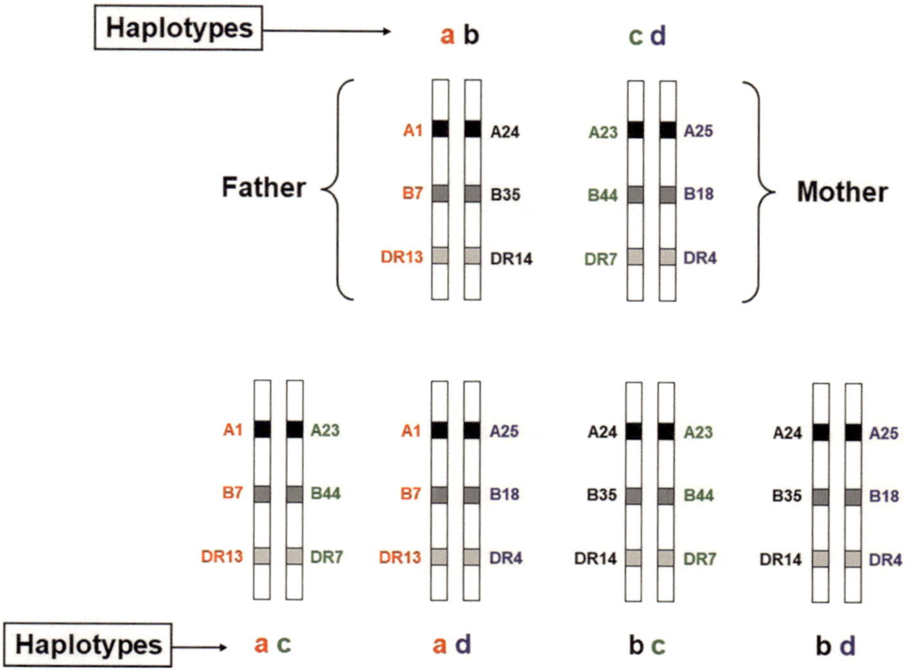

Figure 2.
Inheritance of HLA haplotypes. HLA genes are inherited en block from each parent according to Mendelian laws.

HLA	A	B	C	DRB1	DQB1	Degree of matching
Patient	01:01	07:02	07:02	13:02	06:04	
	25:01	18:01	12:03	04:04	03:02	
Father	01:01	07:02	07:02	13:02	06:04	Haplo-identical
	24:02	35:03	12:03	14:01	05:03	
Mother	23:01	44:03	04:01	07:01	02:02	Haplo-identical
	25:01	18:01	12:03	04:04	03:02	
Sibling 1	01:01	07:02	07:02	13:02	06:04	HLA-identical
	25:01	18:01	12:03	04:04	03:02	
Sibling 2	01:01	07:02	07:02	13:02	06:04	Haplo-identical
	23:01	44:03	04:01	07:01	02:04	
Sibling 3	24:02	35:03	12:03	14:01	05:03	Haplo-identical
	25:01	18:01	12:03	04:04	03:02	
Sibling 4	24:02	35:03	12:03	14:01	05:03	Two haplotype mismatch
	23:01	44:03	04:01	07:01	02:02	

Table 2.
Family segregation analysis for HLA genes.

Several organs and tissues such as kidney, heart, lung, liver, pancreas, skin, bone marrow, hematopoietic stem cells, cornea, and vascularized composite tissue (allografts composed of multiple different tissues such as a hand allograft, which consist of muscle, skin, bone, blood vessels, nerves, and connective tissue) can be transplanted. As mentioned above, all transplanted organs and tissues are called allografts indicating genetic differences between donor and recipient. HLA matching (compatibility) between donor and recipient increases the chance for successful long-term allograft survival [3–5]. For example, if the donor and recipient are not HLA-matched, the recipient's T and B lymphocytes will recognize the foreign donor cells as non-self and will develop an immune response against the allograft resulting in immune-mediated rejection and possible loss of function.

Since HLA mismatches induce the activation of an alloreactive immune response, transplant programs try to match as many as possible donor and recipient HLA molecules. As shown in **Table 3**, better-matched kidney allografts have better survival rates [3–6]. Thus, it is recommended to determine the HLA typing of prospective donor-recipient pairs before transplantation to have a better immunological risk assessment. Optimal HLA matching between donor and recipient prevents allograft rejection in solid organ transplantation (SOT) and prevents both allograft rejection and graft versus host disease (GVHD) in hematopoietic stem cell transplantation (HSCT). Current clinical histocompatibility testing consists of molecular HLA typing and testing for the presence of circulating donor-specific anti-HLA antibodies. The required level of HLA typing resolution varies depending on the transplant type. Potential SOT donors and recipients are typed by means of low-resolution molecular HLA typing techniques. In contrast, potential HSCT donors and recipients require high-resolution HLA typing. Currently, histocompatibility laboratories have replaced serological typing methods with molecular typing methods. These new molecular techniques have significantly improved the histocompatibility laboratory's accuracy and efficiency [7–9].

Several studies have shown that a small but important percentage of kidney allografts are lost during the first-year post-transplantation due to cellular and/or antibody-mediated rejection despite optimal HLA matching (**Table 3**) [3–5]. This observation suggests the presence of mismatches in other non-HLA minor

HLA-A, B, DR mismatches	Survival rates (95% CI)	
	1 year	5 years
0	96.0 (95.4–96.5)	82.8 (81.7–83.8)
1	96.7 (95.7–97.5)	84.0 (82.1–85.6)
2	96.2 (95.6–96.6)	83.1 (82.1–84.1)
3	95.5 (95.1–95.9)	80.8 (80.1–81.5)
4	94.4 (94.0–94.7)	77.8 (77.1–78.5)
5	94.0 (93.7–94.4)	75.9 (75.2–76.5)
6	93.4 (92.9–93.9)	75.7 (74.8–76.6)

Data obtained from the Organ Procurement and Transplantation Network (OPTN): Kidney allografts transplanted between 2008 and 2015 [6].
CI, confidence intervals.

Table 3.
Influence of HLA matching on kidney graft survival.

histocompatibility antigenic systems that also play an important role in allograft rejection [10]. In addition, these data also suggest that low-resolution HLA typing may no longer be considered adequate for SOT patients [11, 12]. Thus, it has been established that better HLA matching results in better allograft survival as well as lower GVHD rates and less need for immunosuppression after transplantation. Hence, HLA typing plays an important role in donor selection and risk assessment in both SOT and HSCT.

2. History

The contribution of the MHC to allograft rejection was first proposed by Bover [13] who observed that skin allografts between identical twins were not rejected like those from genetically different individuals. Then, the MHC genes involved in the allograft rejection process were first described in mice by Gorer [14] Subsequently, Snell [15] used mouse cell lines to further define a locus, which was called H for histocompatibility. Gorer [14] referred to the gene products of locus H as antigens II and the combined term H-2 was consequently used for the mouse MHC. The HLA system was subsequently discovered in the 1950s. Several investigators independently observed that sera from previously transfused individuals and from multiparous women contained antibodies that agglutinated leukocytes [16]. This observation led to the development of serological typing methods that identified a single locus that was subsequently split into two loci: HLA-A and HLA-B. At first, several techniques were used for serological typing but the microlymphocytotoxicity assay became the most widely employed [17–20]. Subsequently, it was observed that, when cultured together in a "mixed lymphocyte culture" (MLC), lymphocytes from unrelated individuals matched at HLA-A and HLA-B loci, showed a robust proliferative response [21, 22]. This observation led to the discovery of an additional locus initially called HLA-D [23–25]; it was subsequently shown that mismatches at HLA-DR and HLA-DQ contributed to the lymphocyte activation observed in the "mixed lymphocyte culture". Soon after, extensive serological studies led to the discovery of HLA-C [26]. Later, HLA-DP, originally called the "secondary B cell" (SB) antigen, was discovered by means of a secondary stimulation assay, called the "primed lymphocyte test" (PLT), that showed the recognition of another HLA molecule different from those recognized in the primary "mixed lymphocyte culture" [27, 28].

3. Genetics and molecular structure

The HLA genetic system is composed of three regions containing genes of different classes (I, II, and III) located on chromosome 6 (**Figure 1**) [29]. The class I region contain genes encoding for HLA-A, B, and C. The class II regions contain genes encoding for HLA-DR, DQ, and DP. Moreover, the class III region is located between the class I and class II and contains genes encoding for molecules involved in immune function that are not targets for allorecognition (**Figure 1**) [29]. Several cytokine genes such as tumor necrosis factor (*TNF*) are found in the class III region. In addition to the main HLA system, the extended MHC system covers 8 Mb and includes the hemochromatosis (*HFE*) gene, the farthest telomeric locus in the system, and the tapasin (*TAPBP*) gene, the farthest centromeric locus of the system. Additionally, several genes that encode for proteins involved in antigen processing and presentation (HLA-DO, *TAP-1*, and subunits of the immune proteasome) are located in the class II region [29].

As mentioned above, HLA class I and class II molecules are expressed differently in different tissues (**Table 1**) [30–33]. Despite the discovery that HLA molecules are the main mediators of transplant rejection, allorecognition is not their main function. HLA molecules are expressed on the surface of cells of the immune system, allowing for cell-cell interactions during the development of an immune response [1, 2]. As shown in **Figure 3**, HLA class I molecules is composed of a long protein of 45 kDa (heavy chain) associated with a smaller protein of 12 kDa called the β2-microglobulin (β2m) encoded by a gene located on chromosome 15. The two chains are associated on the cell surface by non-covalent bonds. The heavy chain has a transmembrane polypeptide, anchoring the complex on the cell surface. HLA class II molecules are composed of two transmembrane proteins, the α chain of 33–35 kDa and the β chain of 26–28 kDa. The two chains associate forming a groove that bind peptide fragments (approximately 15 amino acids long) derived from extracellular proteins that have been processed by the cell (extracellular antigens). In contrast, HLA class I molecules bind peptide fragments (approximately 9 amino acids long) derived from proteins synthesized within the cell (intracellular antigens). HLA class I and class II molecules present the bound peptides to CD8+ and CD4+ T lymphocytes, respectively. The different HLA molecules vary in their efficiency to bind a particular peptide fragment, resulting in a range of immune responses to a given peptide. The different affinity of HLA molecules for a given peptide can affect symptoms of disease, for example, the

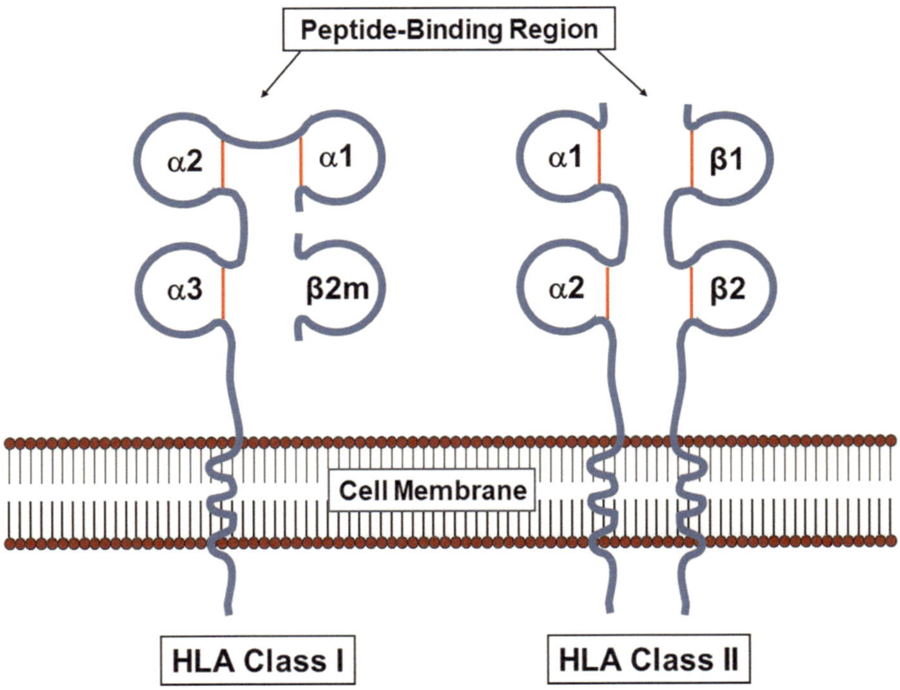

Figure 3.
Structure of the HLA class I and class II molecules. The HLA class I molecules consist of one polymorphic heavy chain (α) associated with a light chain called the β2-microglobulin (β2m). The HLA class II molecules consist of two polymorphic chains (α and β).

susceptibility of individuals carrying a particular HLA genotype to develop acquired immunodeficiency syndrome (AIDS) after human immunodeficiency virus infection (HIV) [34].

4. Polymorphisms

The HLA system is the most polymorphic genetic system of the human genome. As mentioned above, HLA polymorphisms were first defined phenotypically by acceptance or rejection of tissue and/or by reaction with defined alloantibodies (serological typing methods). Subsequently, molecular typing methods showed HLA polymorphisms that range from a single nucleotide change to a loss or gain of an entire genetic region. The identification of HLA polymorphisms was originally performed by serological and cell proliferation assays [18–20, 25]. These assays were successfully used to initially characterize the HLA system; however, despite their extensive application, these have significant limitations in terms of accuracy and reproducibility [35]. Also, alloreactive antisera are usually in limited supply and both serological and cellular HLA typing assays require live cells. Overall, the main limitation of serological HLA typing is its inability to recognize minor polymorphic differences capable to activate CD4+ or CD8+ T lymphocytes.

A general characteristic of the HLA genes is that the distal membrane domains are highly polymorphic, while the proximal membrane domains, the transmembrane, and cytoplasmic domains have very low or no polymorphisms. The heavy chain of the HLA class I molecule is composed of 3 extra-cellular domains and both the α and β chains of the HLA class II molecule contain 2 extracellular domains (**Figure 3**). Of note, the HLA genes, like all eukaryotic genes, contain both coding (exons) and non-coding (introns) regions. The HLA class I genes contain 8 exons while the HLA class II genes contain 6 or 7 exons [1, 2]. The wide application of molecular typing techniques has allowed for the characterization of thousands of HLA alleles [36]. The current

Gene	Serology	Proteins	Alleles	Null alleles
HLA-A	28	4450	7644	397
HLA-B	60	5471	9097	318
HLA-C	10	4218	7609	330
HLA-DRA1	0	5	43	0
HLA-DRB1	21	2203	3389	115
HLA-DRB3	1	334	446	22
HLA-DRB4	1	144	223	25
HLA-DRB5	1	142	187	23
HLA-DQA1	0	244	508	13
HLA-DQB1	9	1455	2330	102
HLA-DPA1	0	233	491	21
HLA-DPB1	6	1325	2221	113

Data obtained from the IPD-IMTG/HLA Database [37–39].

Table 4.
HLA specificities identified by serological and molecular methods (assigned as of September 2022).

number of HLA alleles are shown in **Table 4** and can also be found on the IPD-IMGT/HLA Database [37–39]. As mentioned above, analysis of the nucleotide sequences of the HLA genes indicates that most of the polymorphisms are found in exons 2 and 3 of the HLA class I genes and in exon 2 of the HLA class II genes. These exons encode for the distal membrane domains called the "peptide-binding region" [1, 2, 37, 40, 41]. It has been observed that most nucleotide polymorphisms within the "peptide-binding regions" involve changes that induce a change in the corresponding amino acid sequence (non-synonymous substitutions) and has a high level of correlation with phenotype differences detected by serological and cellular methods [35]. However, serological equivalents are not available for all described alleles [37–39]. Additionally, it is difficult to predict the serological specificities of selected alleles with polymorphisms corresponding to more than one antigenic group [42, 43].

It has been observed that most of the polymorphisms are restricted to some segments of the gene called variable regions. Alleles pairs associated with the same serotype (e.g., A*02:01, A*02:02) differ only by a few nucleotides while distinguishing sequences are observed in alleles of other serotypes, indicating the patchwork nature of HLA polymorphisms. The significant HLA polymorphism probably evolved from the existence of a few allelic lineages followed by short segmental exchanges to increase the number of alleles of a given locus. It appears that most of the HLA polymorphisms was generated by this mechanism. Then, selected natural selection events must have been necessary for new alleles to reach a significant population frequency. Nevertheless, it should be noted that many alleles arose from single-point mutations.

With the understanding of the significant level of HLA polymorphism and the improvement of molecular techniques, several molecular HLA typing methods have been developed [44–46]. These methods have focused on the detection of polymorphisms in exons 2 and 3 of HLA class I genes and in exon 2 of the HLA class II genes. The application of these molecular techniques has resulted in accurate and reproducible HLA typing methods suitable for clinical application [8, 44, 47, 48]. The wide application of these methods has led to the identification of many new alleles that were previously undetectable with the serological and cellular methods [37–39]. The molecular HLA typing methods are widely used and take advantage of the effortlessness of DNA amplification by polymerase chain reaction (PCR). The more widely used molecular HLA typing methods currently used in histocompatibility laboratories are: (1) amplification with sequence-specific primers (SSP), (2) hybridization with sequence-specific oligonucleotide probe hybridization (SSOPH), and (3) direct analysis the DNA sequence (sequence-based typing, SBT) by means of Sanger sequencing or next-generation sequencing (NGS).

5. Nomenclature

A standard nomenclature for serologically defined HLA specificities has been established by the World Health Organization (WHO) nomenclature committee [37–39]. A list of the serologically defined HLA molecules accepted by the WHO is shown in **Table 5** and can also be found on the IPD-IMGT/HLA Database [37–39]. The WHO official nomenclature refers to serologically defined antigens by a number following the gene name, for example, HLA-A2 indicates the HLA-A antigen 2. Subtypes from a broad specificity are followed by the number of the parental antigen in parentheses. For example, HLA-A24(9) indicates the HLA-A antigen 24 derived from

HLA-A	HLA-B	HLA-C	HLA-DR	HLA-DQ	HLA-DP
A1	B5	Cw1	DR1, DR103	DQ1	DPw1
A2, A203, A210	B51(5), B5102, B5103	Cw2	DR2	DQ5(1)	DPw2
A3	B52(5)	Cw3	DR15(2)	DQ6(1)	DPw3
A9	B7, B703	Cw9(w3)	DR15(2)	DQ2	DPw4
A23(9)	B8	Cw10(w3)	DR3	DQ3	DPw5
A24(9), A2403	B12	Cw4	DR17(3)	DQ7(3)	DPw6
A10	B44(12)	Cw5	DR18(3)	DQ8(3)	
A25(10)	B45(12)	Cw6	DR4	DQ9(3)	
A26(10)	B13	Cw7	DR5	DQ4	
A34(10)	B14	Cw8	DR11(5)		
A66(10)	B64(14)		DR12(5)		
A11	B65(14)		DR6		
A19	B15		DR13(6)		
A29(19)	B62(15)		DR14(6), DR1403, DR1404		
A30(19)	B63(15)		DR7		
A31(19)	B75(15)		DR8		
A32(19)	B76(15)		DR9		
A33(19)	B77(15)		DR10		
A74(19)	B16				
A28	B38(16)		DR51		
A68(28)	B39(16), B3901, B3902		DR52		
A69(28)	B17		DR53		
A36	B57(17)				
A43	B58(17)				
A80	B18				
	B21				
	B49(21)				
	B50(21)				
	B22				
	B54(22)				
	B55(22)				
	B56(22)				
	B27, B2708				
	B35				
	B37				
	B40, B4005				
	B60(40)				
	B61(40)				

HLA-A	HLA-B	HLA-C	HLA-DR	HLA-DQ	HLA-DP
	B41				
	B42				
	B46				
	B47				
	B48				
	B53				
	B59				
	B67				
	B70				
	B71(70)				
	B72(70)				
	B73				
	B78				
	B81				
	B82				
	Bw4*				
	Bw6*				

Broad antigen specificities are listed in parentheses.
Associated antigens (e.g., A2, A203, A210) are listed together.
**Bw4-associated specificities: A23, A24, A2403, A25, A32, B13, B27, B37, B38, B44, B47, B49, B51, B5102, B5103, B52, B53, B57, B58, B59, B63, B77.*
***Bw6-associated specificities: B7, B703, B8, B18, B2708, B35, B39, B3901, B3902, B4005, B41, B42, B45, B46, B48, B50, B54, B55, B56, B60, B61, B62, B64, B65, B67, B71, B72, B73, B75, B76, B78, B81, B82.*
Data obtained from the IPD-IMGT/HLA Database [37–39].

Table 5.
Serologically recognized HLA specificities.

the parental HLA-A antigen 9. In this regard, the derived antigens are called "split" specificities. As mentioned above, the current numbers of HLA specificities identified by serological and molecular methods are shown in **Table 4** and can also be found on the IPD-IMGT/HLA Database [37–39].

With the introduction of molecular HLA typing methods in the 1980s, HLA typing at the DNA level required nomenclature for specific DNA sequences [37–39]. A revised nomenclature is used for designating alleles (**Figure 4**) [49, 50]. Each HLA allele has a unique number corresponding to up to four sets of digits separated by colons. All alleles receive at least a four-digit name, which corresponds to the first two sets of digits; longer names are assigned only when necessary. The gene name, such as HLA-A, is followed by an asterisk and the allele family number (which often corresponds to the serological typing); for example, A*02 (**Figure 4**). The second set of digits is used to list the subtypes of the allele family, the numbers have been assigned in the order in which DNA sequences have been found. Alleles whose numbers differ in the two sets of digits must differ in one or more nucleotide(s) that change the amino acid sequence of the encoded protein; for example, A*02:01 and A*02:02 (**Figure 4**). Alleles that differ by a synonymous nucleotide substitution in the exons

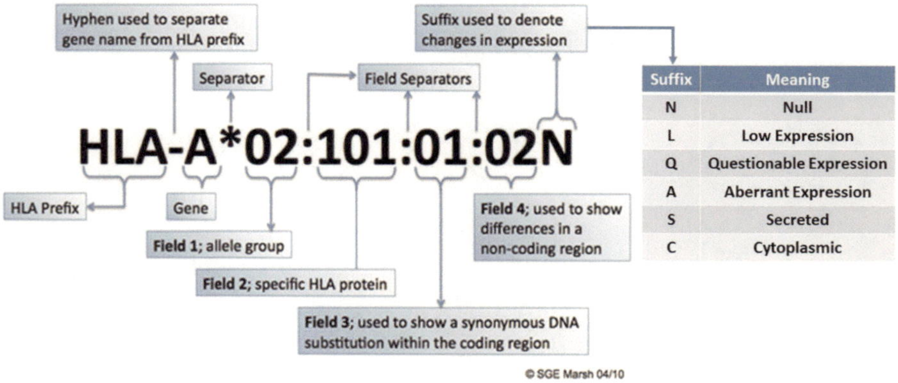

Figure 4.
HLA nomenclature. The gene is indicated after the HLA prefix and hyphen. Then an asterisk is used as a separator before the first field, which is represented by two digits. A colon is used to separate each of the fields, with up to four fields listed. The second field indicates the specific HLA protein, while the third field is used to describe a nucleotide change within the coding region that does not change the encoded amino acid. The fourth field describes changes in non-coding regions. A suffix may be utilized to indicate changes in expression, and the potential suffixes are listed in the table. The figure is courtesy of Professor Steven G.E. Marsh, Anthony Nolan Research Institute, London, United Kingdom [38].

(changes in the DNA sequence that do not change the amino acid sequence), also called silent mutations or non-coding substitutions, are distinguished using the third set of digits; for example, A*02:01:01 and A*02:01:02 (**Figure 4**). Alleles that differ by a nucleotide substitution in the introns, or in the 5′ or 3′ untranslated regions that flank the exons and introns, are distinguished using the fourth set of digits; for example, A*02:01:01:01 and A*02:01:01:03 (**Figure 4**).

In addition to the unique allele number, there are additional suffixes that may be added to an allele to indicate its expression status (**Figure 4**) [37–39]. Alleles that have been shown not to be expressed have been given the suffix "N" for "Null"; for example, HLA-B*13:07 N is not expressed due to a 15-base pair deletion in exon 2. "Null" alleles can also be the result of premature stop codon mutations, nonsense, frameshift, and splice site. In addition, the suffix "L" indicates a "Low" cell surface expression as compared to normal expression levels. The suffix "Q" indicates a "Questionable" cell surface expression, given that the allele carries a mutation that has been shown to affect normal expression levels. The suffix "A" indicates an "Aberrant" cell surface expression, specifically where there is doubt as to whether the allele product has any cell surface expression. The suffix "S" indicates an allele product that is a "Secreted" molecule with no cell surface expression. The suffix "C" indicates an allele product that is present in the "Cytoplasm" with no cell surface expression. A list of the "Null" and alternatively expressed alleles can be found on the IPD-IMGT/HLA Database [37–39].

HLA specificities can be identified to varying degrees of resolution depending on the methodology used [8]. In this regard, the definitions for low, high, and allelic resolution typing were compiled by the "Harmonization of Histocompatibility Typing Terms Working Group", to define a language for histocompatibility laboratories to report HLA typing results [40, 41]. An allelic-resolution HLA typing result is consistent with a single allele as defined by the WHO HLA Nomenclature Report [37–39]. An allele is defined as a unique nucleotide sequence for a gene defined using all the digits in a current allele name; for example, A*02:01:01:01 (**Figure 5**). A high-resolution HLA typing result is defined as a group of alleles that encode the same protein sequence for the region of the HLA molecule called the "peptide-binding

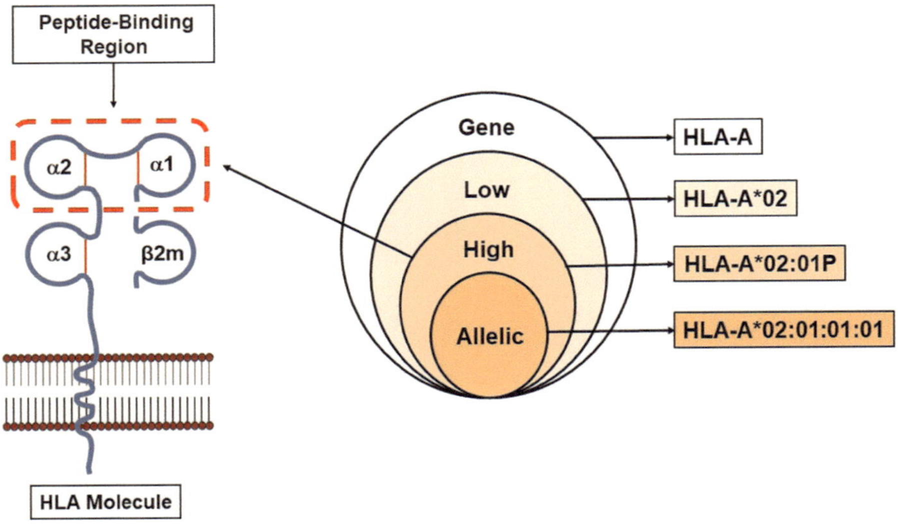

Figure 5.
HLA typing resolution. The Venn diagram illustrates increasing levels of HLA typing resolution. The figure on the left shows the "peptide-binding region" of an HLA class I molecule. High-resolution HLA typing defines the specific DNA sequence of the 'peptide-binding region". Allelic resolution defines a single allele as defined by a unique DNA sequence for the HLA gene. Adapted from Nunes et al. [40, 41].

region" (**Figure 5**). The "peptide-binding region" includes domains 1 and 2 of the HLA class I α chain (encoded by exons 2 and 3) and domain 1 of the HLA class II α and β chains (encoded by exon 2) [1, 2]. This group of alleles is designated by an upper case "P" which follows the 2-field allele designation of the lowest numbered allele in the group; for example, A*02:01P (**Figure 5**). A low-resolution HLA typing result is defined as the digits composing the first field in the DNA-based nomenclature; for example, A*02 (**Figure 5**). If the resolution corresponds to a serologic equivalent, this typing result should also be called low-resolution; for example, A2 [40, 41].

6. Typing techniques

6.1 Serological typing

As mentioned above, histocompatibility testing was traditionally performed using serological methods utilizing (1) sera with known HLA specificity to identify HLA antigens and (2) using cells with known HLA antigens to identify anti-HLA antibodies in patient sera [17–20]. Although serological methods yield only low-resolution HLA typing results, there are some advantages to these methods; serological methods are relatively rapid and will reveal immunologically relevant epitopes. Also, serological methods can be used to confirm "Null" alleles detected by molecular methods.

6.2 Molecular typing

Several molecular HLA typing methods are now widely available to the histocompatibility laboratory [8, 44, 46, 51]. These methods can be broadly classified into two

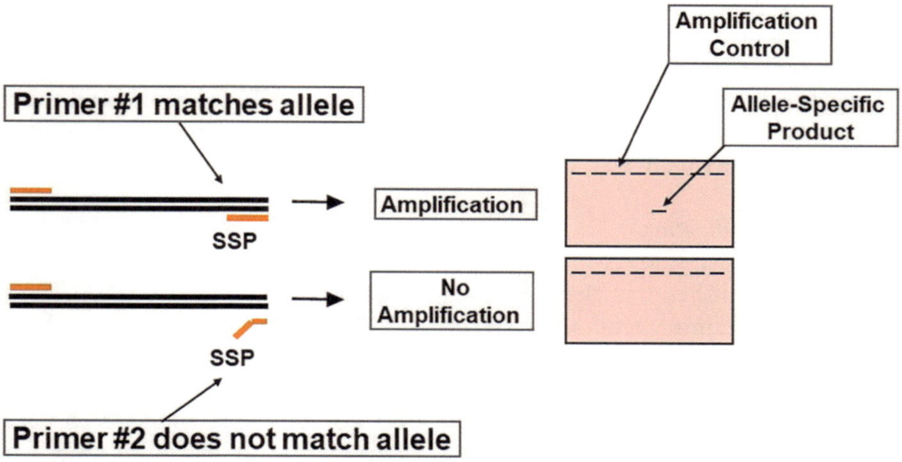

Figure 6.
Principle of sequence-specific primers. HLA alleles are amplified by PCR using SSP. The PCR products are then detected by agarose gel electrophoresis or real-time PCR amplification plots. An amplification control is included with each reaction to detect false negative results due to amplification failure.

categories: (1) methods in which the polymorphisms are identified directly by the PCR process, without further steps, for example SSP; and (2) methods that generate PCR products containing polymorphisms that can be identified by a secondary technique, for example SSOPH, Sanger sequencing, and NGS.

The use of specific techniques depends on the histocompatibility laboratory's requirements for HLA typing resolution and volumes. The different methods have different requirements in terms of skills of testing personnel, instruments, and turnaround time. Currently, most histocompatibility laboratories use a combination of HLA typing techniques to obtain the desired level of resolution. Currently, low-resolution HLA typing is sufficient for SOT while HSCT requires allelic- or high-resolution HLA typing depending on the internal policies of the transplant program (**Figure 6**) [52]. In recent years, it has been shown that alloantibodies may also recognize allelic differences; therefore, in some cases, high-resolution HLA typing may also be required for SOT [7, 11].

6.2.1 Steps for molecular typing

Molecular HLA typing techniques involve three general steps: (1) extraction of genomic DNA, (2) amplification of segments of the gene of interest and (3) detection of the sequence polymorphisms that define the alleles.

6.2.1.1 DNA extraction

Genomic DNA is usually extracted in the histocompatibility laboratory from peripheral blood leukocytes. Only a few micrograms of genomic DNA are sufficient to perform any of the molecular HLA typing methods described below. The most widely used DNA extraction method in the histocompatibility laboratory is the "salting-out" method. This is a simple and non-toxic DNA extraction method developed by Miller et al. [53] During this method, proteinase K and RNase are added to the sample after cell lysis. Proteinase K is used to digest proteins and remove contamination from the sample. In this regard, addition of proteinase K rapidly inactivates nucleases that might otherwise degrade the DNA during purification. Then, saturated NaCl is added to precipitate the proteins. Then, the sample is centrifuged, and the DNA, present in the supernatant, is transferred to a second tube to be centrifuged again, and subsequently washed with 70% ethanol. High yields of high-quality DNA are obtained with this method in a relatively fast and inexpensive manner.

6.2.1.2 DNA Amplification

Once it is established that there is enough quantity of high-quality DNA. The PCR mixture, containing genomic DNA, primers, Taq DNA polymerase, and four types of deoxynucleotide triphosphate (dNTP) (adenine, dATP; cytosine, dCTP; guanine, dGTP; and thymidine, dTTP), is prepared. Then, the PCR mixture is subjected to repeated cycles of heating to 94–96°C for double-stranded DNA denaturation; cooling down to the corresponding temperature for primer annealing and lastly, warming up to 72°C for an optimal Taq DNA polymerase activity to integrate the complementary dNTP to the single-stranded DNA. After the first amplification cycle, the DNA copies serve as templates to allow exponential growth of the PCR product. Because both DNA strands need to be amplified, the primers for those strands should be designed to include intervening segments between the two strands. To ensure the optimal efficiency of the PCR reaction, the number of cycles and the duration of incubation periods at each temperature should be based on the length of the DNA fragments and GC content from both the DNA fragments and the primers.

6.2.1.3 Detection of sequence polymorphisms

6.2.1.3.1 Sequence-specific primers

SSP is a rapid method that uses sets of primer pairs to amplify genomic DNA. The efficiency of the amplification reaction is controlled by the primers that amplify conserved sequences of a selected gene. The 3′ end of the primer must match the template for recognition by the Taq DNA polymerase. By designing primers with polymorphic sequences at the 3′ end, successful amplification can be used to type specific HLA alleles (allelic-resolution) or a group of alleles (low- or high-resolution). The HLA typing results are interpreted by analyzing the amplification pattern, detected on an agarose gel electrophoresis (**Figure 6**).

SSP reactions can be set up in a 96-well plate format with different allele-specific primer sets in each well. Each PCR mixture contains sequence-specific primers and a set of amplification control primers. The amplification control primers should yield a PCR product for every specimen (except the negative control). The sequence-specific primers should only yield a product if the specimen has the HLA specificity matching the primers' sequences. The amplification control primers are designed to yield a PCR product of different size from the product of the HLA-specific primers. The two PCR products are then resolved by agarose gel electrophoresis or real-time PCR. Specimens will yield two PCR products (amplification control and HLA-specific product) only from those PCR reactions containing primers matching the sample's HLA specificity (**Figure 6**). PCR reactions containing primers that do not match the sample's HLA specificity should only show the amplification control. This method can easily be implemented, HLA typing can be performed in 4–5 h, and it is adequate for low-volume histocompatibility laboratories.

6.2.1.3.2 Sequence-specific oligonucleotide probe hybridization

Specific PCR amplification of an HLA locus and the subsequent probing of the PCR product with probes immobilized on an array of Luminex® microbeads is the method known as SSOPH. The Luminex® assay uses color-coded microbeads of defined spectral properties conjugated to sequence-specific probes and incubated with the samples in a 96-well microplate (**Figure 7**). The Luminex® technology is a bead-based assay that allows for multiplex detection of up to 100 analytes simultaneously. For this method, the HLA locus is first amplified using primers flanking the polymorphic sequences and labeled with biotin at the 5′ end. The probes are short (approximately 20 nucleotides long) single-stranded DNA fragments designed to hybridize to a specific HLA specificity (**Figure 7**). The probe sequences, based on sequences found on the IPD-IMGT/HLA Database, are aligned so that the polymorphic sequences are in the middle of the probe's sequence [37–39]. The amplified biotin-labeled DNA binds to the probes over during a pre-selected period of incubation (**Figure 7**). After the first incubation, the microbeads are washed, and a second incubation follows in the presence of a biotin-specific molecule, streptavidin, labeled with a fluorescent dye, phycoerythrin (**Figure 7**). After the second incubation, the microbeads are washed and ready to analyze. The microbead array is then classified as phycoerythrin-positive or negative in a Luminex® flow cytometry instrument. Different patterns of phycoerythrin-positive microbeads define the different HLA alleles (**Figure 7**). The number of probes used in the assay depends on the desired HLA typing resolution. Of note, because of multiple recombination events generating polymorphism in the HLA system, many of the sequences in the polymorphic regions are not allele-specific, i.e.,

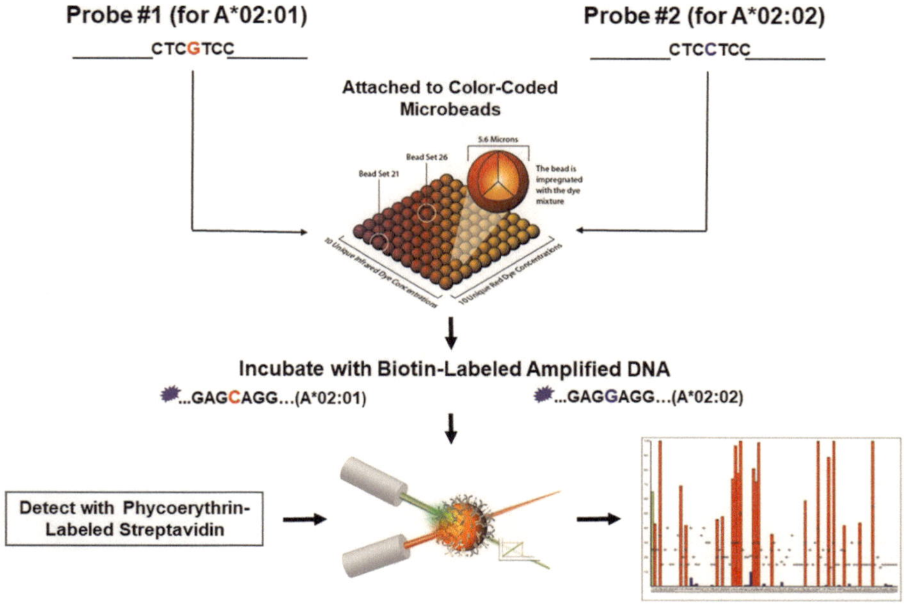

Figure 7.
Principle of sequence-specific oligonucleotide probe hybridization. HLA genetic regions are amplified by PCR using generic primers covalently bound to biotin at the 5′ end. The amplified and labeled DNA bound to the probes immobilized on Luminex® microbeads is then detected with a biotin-specific molecule, streptavidin, conjugated with phycoerythrin. The microbead array is then classified as phycoerythrin-positive or negative by a Luminex® flow cytometry instrument. Different patterns of phycoerythrin-positive and negative microbeads define specific HLA alleles.

several sequences are shared by many HLA alleles. Therefore, since selected probes detect multiple HLA specificities, hybridization profiles are usually highly complex, and a software is needed for accurate interpretation of the results (**Figure** 7).

6.2.1.3.3 Sequence-based typing by sanger sequencing

The most accurate method for HLA typing is the direct identification of the DNA sequence. Until recently, the most widely used sequence-based HLA typing method was the "chain termination method" that is performed by the random incorporation of four types of dideoxynucleotide triphosphate (ddNTP) (adenine, ddATP; cytosine, ddCTP; guanine, ddGTP; and thymidine, ddTTP) labeled with four different fluorescent dyes (**Figure 8**) [54]. This method is named after the inventor of this ground-breaking technology, Dr. Frederick Sanger, who developed this method in the 1970s [54]. In the second step, the chain-terminated DNA fragments are separated by size via gel electrophoresis; the DNA samples are loaded into one end of a gel matrix, and an electric current is applied; DNA is negatively charged, so the DNA fragments are pulled toward the positive electrode on the opposite side of the gel. Because all DNA fragments have the same charge per unit of mass, the speed at which the DNA fragments move will be determined only by size. The smaller a DNA fragment is, the less friction it will experience as it moves through the gel, and the faster it will move. As a result, the labeled dideoxynucleotides attached to the DNA fragments will be arranged from smallest to largest, and they will be detected first reading the gel from bottom to top (**Figure 8**). The last step involves reading the gel to determine the DNA sequence. Since DNA polymerase only synthesizes DNA in the 5′–3′ direction starting at a provided primer, each terminal ddNTP will correspond to a specific nucleotide in the original DNA sequence (e.g., the shortest DNA fragment must terminate at the first nucleotide from the 5′ end, the second-shortest DNA fragment must terminate at the second nucleotide from the 5′ end, etc.). Therefore, by reading the gel bands from the smallest to largest DNA fragment, the 5′–3′ sequence of the original DNA strand can be determined. A computer reads each band of the gel, in order, using fluorescence to call the identity of each terminal ddNTP. Briefly, a laser excites the

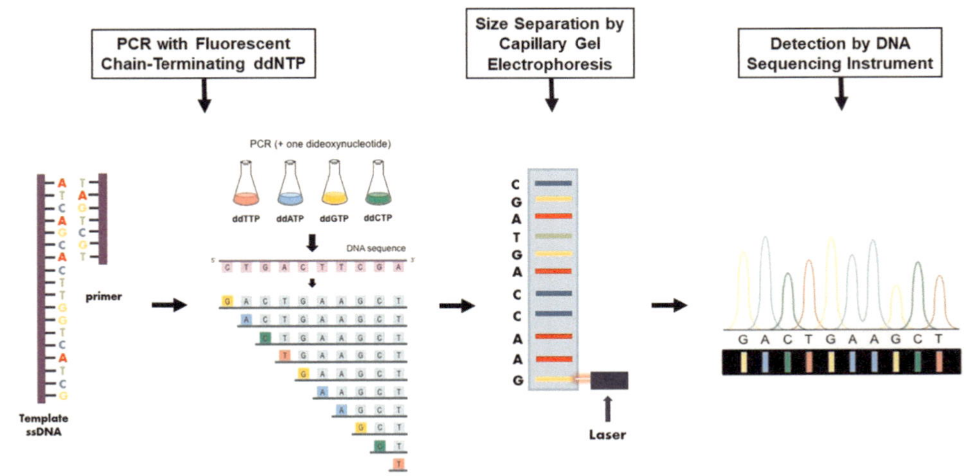

Figure 8.
Principle of sequence-based typing by Sanger sequencing. HLA genetic regions are amplified by PCR using locus-specific primers. The PCR products are then purified from unused PCR reaction components, sequencing reactions are performed using forward and reverse sequencing primers and these reactions are loaded onto the automated DNA sequencer to detect the nucleotide sequence of the targeted genes.

fluorescent terminal ddNTP in each band, and a computer detects the resulting light emitted. Because each of the four ddNTP is labeled with a different fluorescent dye, the light emitted can be directly tied to the identity of the terminal ddNTP. The output is called a chromatogram, which shows the fluorescent peak of each nucleotide along the length of the template DNA (**Figure 8**).

When a single PCR reaction is performed in a particular sample, simultaneous amplification and sequencing of both alleles are obtained and heterozygous nucleotide assignments are observed at positions where both alleles have different nucleotides. Some heterozygous genotypes displaying the same sequencing pattern result in ambiguous (alternative) sequencing results. Performing additional tests targeting only one of the possible alleles, either by additional sequencing primers, SSP, and/or SSOPH usually resolves these ambiguous results, enabling the separation of the two alleles in heterozygous samples.

6.2.1.3.4 Sequence-based typing by next-generation sequencing

HLA typing by Sanger sequencing is restricted by its low throughput and high cost. In addition, HLA typing ambiguities, resulting from the inability to phase heterozygous nucleotide positions, requires time-consuming follow-up testing to resolve the ambiguous HLA typing results. Short-read NGS, also called second-generation sequencing has begun to alleviate these disadvantages for HLA typing. The common element of short-read NGS technologies is massive sequencing of short (250–800 basepairs long), clonally amplified DNA fragments sequenced in parallel [55]. Several NGS-based HLA typing kits are commercially available, all of which have achieved similar accuracy and straightforward multiplexing workflows covering the clinically relevant HLA class I and class II genes.

The current NGS workflow includes the following steps: DNA isolation, library preparation, and sequencing (**Figure 9**). DNA can be extracted from anticoagulated whole blood or buccal swabs utilizing commercial kits. High quality DNA, free of biological and chemical contaminants, is crucial for sequencing success. Due to limitations in the sensitivity of standard spectrophotometry, fluorometers are advantageous for analyzing DNA concentration, but cannot provide absorbance ratios. A necessity for NGS is a good quality library. During library preparation DNA templates are obtained which are compatible with the sequencing instrument utilized for NGS. Library enrichment protocols in general depend on one of the following methods: short-range PCR, long range PCR, or hybrid capture. For PCR-based library enrichment protocols, multiplex primers are used to amplify the HLA loci of interest. Short-range PCR-based enrichment for HLA typing focuses on the amplification of exons encoding the "peptide-binding region" and has the advantage of faster sequencing and higher depth of coverage for the enriched genomic regions. However, the possible loss of phasing over longer segments of DNA has been observed with this methodology resulting in typing ambiguities seen with Sanger sequencing [56]. Long-range PCR substitutes a longer library preparation time to accomplish amplification of whole HLA genes including introns, upstream and downstream flanking sequences, and in some cases, untranslated genomic regions. Thus, long-range PCR improved issues of HLA typing ambiguities observed with short amplicon sequencing. It should be noted, however, that long-range PCR-based protocols have the potential to show allele imbalance or dropout because of where the primers anneal and the sequence composition in these genomic regions. Consequently, it has been recommended that homozygous HLA alleles or typing results suspected to be affected by allele dropout should

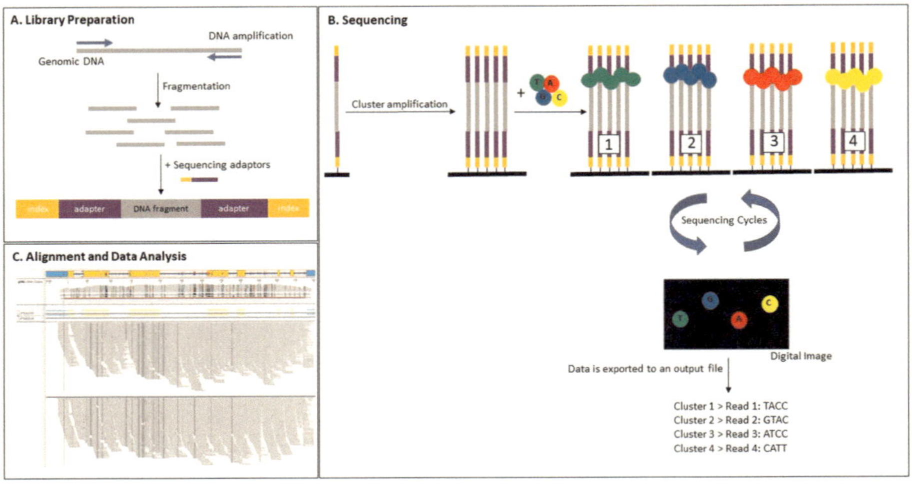

Figure 9.
Principle of sequence-based typing by next-generation sequencing. (A) The sequencing library is prepared by fragmenting the amplified genomic DNA and ligating specialized adapters with indices to both fragment ends. (B) The sequencing library is loaded into a flow cell and the fragments are hybridized to the flow cell surface. Each bound fragment is amplified into a clonal cluster. Sequencing reagents, including fluorescent-dye labeled nucleotides, are added and the first nucleotide is incorporated. The flow cell is imaged and the emitted light from each fragment cluster is recorded. The emission wavelength and intensity are used to identify the incorporated nucleotide. This cycle is repeated several times. (C) Reads are aligned to a reference sequence from the IPD-IMGT/HLA Database [37–39]. After alignment, differences between the reference genome and the newly sequenced reads can be identified.

be typed by another molecular methodology to ensure correct results. The most recent library enrichment methodology utilizes HLA locus complementary oligonucleotide probes which hybridize to the locus of interest. These probes are bound to magnetic beads, which allow for the selection and enrichment of HLA genes during the library preparation step [57]. The benefit of hybrid capture-based library enrichment is that there is no need for long-range PCR, therefore saving PCR time and avoiding PCR amplification errors. However, this methodology has an intrinsic bias in capturing AT/GC-rich DNA regions, and highly repetitive sequences may be underrepresented. Since HLA class II genes carry large intronic areas consisting of repeat elements, it is challenging to design oligonucleotide probes with sufficient specificity to capture these sequences. Consequently, probe-based assays only cover the full exons of HLA class II genes. The lack of coverage in introns can result in typing ambiguities for HLA class II genes, but mainly in the 4th field of typing.

Library preparation includes the additional steps of fragmentation, end-repair, adaptor ligation, and size selection. Fragmentation shears DNA into the ideal NGS platform-specific size range. DNA fragmentation can be achieved by sonication, transposase "tagmentation", or heat treatment with divalent metal cations. Importantly, shorter DNA fragments offer high-quality sequencing data, whereas longer fragments provide distal phasing information. Subsequent end-repair of the DNA fragments prepares the sequencing library for adaptor ligation. Adaptors comprise NGS platform-specific sequences for DNA fragment recognition by the instrument and as well as a unique sequence, called an index or barcode, for labeling of an individual sample thereby allowing several patient samples to be pooled and sequenced concurrently in a single run (multiplex sequencing). Following adaptor ligation, a size selection step enriches DNA fragments within a defined size range and removes contaminations to improve sequencing quality and efficiency. Size selection can be accomplished by utilizing bead-based or electrophoretic-based methodologies. Bead-based approaches allow for simultaneously concentrating the sequencing library, while electrophoretic-based methods enhance precision. Alternatively, DNA fragments can be generated by utilizing an on-bead "tagmentation" protocol, which combines the library preparation steps of DNA normalization, fragmentation, and size selection. Following "tagmentation", a PCR step is performed to incorporate the sequencing adapters and barcodes for patient sample identification. This workflow is straightforward and fast, allowing sequencing libraries to be generated in less than 90 min, with less than 15 min of hands-on time [51].

Since NGS produces substantial amounts of data, efficient bioinformatics data analysis, and data management are critical for successful implementation of NGS in the HLA laboratory. Primary sequencing data analysis is performed by the instrument and involves base-calling for each clonally amplified DNA fragment. Quality control procedures including read filtering and trimming, also take place during this step. Sequencing data is saved together with the quality matrix as a FASTQ file. For HLA genotyping, specific software programs are commercially available for the final analysis steps. As a result of the highly polymorphic nature of HLA genes, alignment to the human reference genome is inefficient for precisely determining the HLA alleles present within a patient sample, and instead depend on alignment to the IPD-IMGT/HLA Database, which contains the sequences of all currently characterized HLA alleles [37–39]. Another quality indicator at this step is coverage, which includes depth of coverage (number of times a base is sequenced) and breadth of coverage (percentage of a reference genome covered). Importantly, coverage may not be consistent throughout the amplicon, and lack of coverage in key regions such as exons may affect the accuracy of the HLA typing result. Commercially available HLA software analysis programs designed for HLA-typing, typically have built-in filters to define the minimum coverage needed for accurate HLA genotyping, although some situations may allow for a lower threshold, such as when the polymorphisms of two alleles of a locus are phased, or when the region with low coverage, such as introns and untranslated genomic regions, does not affect the HLA typing. An important concern for HLA typing data analysis is the evaluation of adequate allele balance to detect issues related to allele dropout, caused either by preferential amplification due to technical issues or the patient's disease state whereby one of the two alleles has been eradicated (loss of heterozygosity).

More recently, novel long-read methods, also known as third-generation sequencing technologies have been established, which generate sequences >10 kb directly from genomic DNA. While early iterations of these methods were challenged with inaccuracies, latest improvements have permitted much higher accuracy and offer the advantage to sequence large DNA fragments in comparison to NGS. Long-read sequencing is particularly advantageous for HLA typing because it would facilitate complete phasing of alleles and mitigate further the challenge of HLA typing ambiguities [58].

7. Concluding remarks

It has been established that HLA incompatibility is an important risk factor for early graft loss after SOT and for development of GVHD and lack of graft engraftment after HSCT. As a result of improvement in immunosuppressive protocols, surgical techniques, and the management of peri- and post-operative clinical complications, short-term allograft survival has greatly increased. However, long-term allograft survival is still hampered by the development of chronic rejection [59, 60]. The current hypothesis is that chronic allograft rejection is the result of a common lesion in which different inflammatory triggers such as rejection episodes, ischemia-reperfusion injury, and infection will lead to a similar histological and clinical outcome. Evidently, the process of chronic rejection is mainly the combined effect of an indolent immune response as well as non-immunologic factors [10, 59, 61–66]. In this regard, a growing body of evidence has demonstrated that chronic rejection is mainly due to the immune response developed against mismatched HLA molecules of the allograft. This is supported by the observation that the main risk factors for the development of chronic rejection after SOT are HLA mismatches between donor and recipient and the severity and frequency of humoral and/or cellular acute rejection episodes. Moreover, the main risk factor for GVHD after HSCT is the presence of HLA mismatches between donor and recipient.

Several studies have shown that the development of *de novo* donor-specific anti-HLA antibodies is associated with the development of chronic rejection and allograft disfunction after SOT [62, 67–70]. Related studies have also shown that pre-existing donor-specific anti-HLA antibodies are associated with early allograft dysfunction after SOT [62, 68, 69]. It was previously suggested that the development of *de novo* donor-specific anti-HLA antibodies was as an epiphenomenon to the activation of the cellular immune response during the development of chronic rejection. However, there is now compelling evidence that donor-specific anti-HLA antibodies play an important role in the chronic rejection process. In this regard, several studies have demonstrated that anti-HLA antibodies induced intracellular signal transduction in both endothelial and epithelial cells resulting in cellular activation, proliferation, and apoptosis as well as the production of cytokines, chemokines, and fibrogenic growth factors [65, 67, 71].

Chronic rejection after SOT and GVHD after HSCT are the result of a failure of current immunosuppressive protocols to control the alloreactive immune response primarily to mismatched HLA molecules. In this regard, it has been suggested that CD4+ T lymphocytes activated through the indirect allorecognition pathway are less responsive to conventional immunosuppression as compared to those activated through the direct allorecognition pathway [72]. To better prevent chronic allograft rejection and GVHD, immunosuppressive protocols that better block the indirect allorecognition pathway are needed. As organ donation rates plateau, extending the life of transplanted organs is of paramount importance. This objective can only be accomplished through better understanding of the immunological processes that occur during the chronic allograft rejection and GVHD.

Author details

Andrés Jaramillo* and Katrin Hacke
Department of Laboratory Medicine and Pathology, Mayo Clinic, Phoenix, Arizona, USA

*Address all correspondence to: jaramillo.andres@mayo.edu

References

[1] Klein J, Sato A. The HLA system. First of two parts. The New England Journal of Medicine. 2000;**343**(10):702-709

[2] Klein J, Sato A. The HLA system. Second of two parts. The New England Journal of Medicine. 2000;**343**(11): 782-786

[3] Kim JJ, Fuggle SV, Marks SD. Does HLA matching matter in the modern era of renal transplantation? Pediatric Nephrology. 2021;**36**(1):31-40

[4] Yacoub R, Nadkarni GN, Cravedi P, He JC, Delaney VB, Kent R, et al. Analysis of OPTN/UNOS registry suggests the number of HLA matches and not mismatches is a stronger independent predictor of kidney transplant survival. Kidney International. 2018;**93**(2):482-490

[5] Opelz G, Wujciak T, Dohler B, Scherer S, Mytilineos J. HLA compatibility and organ transplant survival. Collaborative Transplant Study. Reviews in Immunogenetics. 1999;**1**(3): 334-342

[6] Organ Procurement and Transplantation Network (OPTN) Database. Available from: https://optn.transplant.hrsa.gov/

[7] Huang Y, Dinh A, Heron S, Gasiewski A, Kneib C, Mehler H, et al. Assessing the utilization of high-resolution 2-field HLA typing in solid organ transplantation. American Journal of Transplantation. 2019;**19**(7): 1955-1963

[8] Smith AG, Pereira S, Jaramillo A, Stoll ST, Khan FM, Berka N, et al. Comparison of sequence-specific oligonucleotide probe vs next generation sequencing for HLA-A, B, C, DRB1, DRB3/B4/B5, DQA1, DQB1, DPA1, and DPB1 typing: Toward single-pass high-resolution HLA typing in support of solid organ and hematopoietic cell transplant programs. Hla. 2019;**94**(3):296-306

[9] Marino SG, Jaramillo A, Fernández-Viña MA. The human major histocompatibility complex and DNA-based typing of human leukocyte antigens for transplantation. In: O'Gorman MRG, Donnenberg AD, editors. Handbook of Human Immunology. 2nd ed. New York, NY, USA: CRC Press - Taylor & Francis Group; 2008. pp. 541-564

[10] Lefaucheur C, Louis K, Philippe A, Loupy A, Coates PT. The emerging field of non-human leukocyte antigen antibodies in transplant medicine and beyond. Kidney International. 2021; **100**(4):787-798

[11] Senev A, Emonds MP, Van Sandt V, Lerut E, Coemans M, Sprangers B, et al. Clinical importance of extended second field high-resolution HLA genotyping for kidney transplantation. American Journal of Transplantation. 2020;**20**(12): 3367-3378

[12] Zavyalova D, Abraha J, Rao P, Morris GP. Incidence and impact of allele-specific anti-HLA antibodies and high-resolution HLA genotyping on assessing immunologic compatibility. Human Immunology. 2021;**82**(3): 147-154

[13] Bover KH. Homoisotransplantation von Epidermis bei eineiigen Zwillingen. Beiträge zur Klinischen Chirurgie. 1927; **141**:442-447

[14] Gorer PA. The genetic and antigenic basis for tumor transplantation. The

Journal of Pathology and Bacteriology. 1937;**44**:691-697

[15] Snell GD. Methods for the study of histocompatibility genes. Journal of Genetics. 1948;**49**(2):87-108

[16] Payne R. The development and persistence of leukoagglutinins in parous women. Blood. 1962;**19**:411-424

[17] Terasaki PI, Mandell M, Vandewater J, Edgington TS. Human blood lymphocyte cytotoxicity reactions with allogenic antisera. Annals of the New York Academy of Sciences. 1964; **120**:322-334

[18] Terasaki PI, McClelland JD. Microdroplet assay of human serum cytotoxins. Nature. 1964;**204**: 998-1000

[19] Terasaki PI, Mickey MR. Histocompatibility-transplant correlation, reproducibility, and new matching methods. Transplantation Proceedings. 1971;**3**(2):157-171

[20] Terasaki PI, Rich NE. Quantitative determination of antibody and complement directed against lymphocytes. Journal of Immunology. 1964;**92**:128-138

[21] Bach FH, Amos DB. Hu-1: Major histocompatibility locus in man. Science. 1967;**156**(3781):1506-1508

[22] Bach FH, Kisken WA. Predictive value of results of mixed leukocyte cultures for skin allograft survival in man. Transplantation. 1967;**5**(4, Suppl): 1046-1052

[23] Williams RC Jr, Emmons JD, Yunis EJ. Studies of human sera with cytotoxic activity. The Journal of Clinical Investigation. 1971;**50**(7): 1514-1524

[24] Yunis EJ, Amos DB. Three closely linked genetic systems relevant to transplantation. Proceedings of the National Academy of Sciences of the United States of America. 1971;**68**(12): 3031-3035

[25] Yunis EJ, Plate JM, Ward FE, Seigler HF, Amos DB. Anomalous MLR responsiveness among siblings. Transplantation Proceedings. 1971;**3**(1): 118-120

[26] Thorsby E, Sandberg L, Lindholm A, Kissmeyer-Nielsen F. The HL-A system: Evidence of a third sub-locus. Scandinavian Journal of Haematology. 1970;**7**(3):195-200

[27] Shaw S, Duquesnoy RJ, Smith PL. Population studies of the HLA-linked SB antigens. Immunogenetics. 1981; **14**(1–2):153-162

[28] Shaw S, Johnson AH, Shearer GM. Evidence for a new segregant series of B cell antigens that are encoded in the HLA-D region and that stimulate secondary allogenic proliferative and cytotoxic responses. The Journal of Experimental Medicine. 1980;**152**(3): 565-580

[29] Shiina T, Hosomichi K, Inoko H, Kulski JK. The HLA genomic loci map: Expression, interaction, diversity and disease. Journal of Human Genetics. 2009;**54**(1):15-39

[30] Boegel S, Lower M, Bukur T, Sorn P, Castle JC, Sahin U. HLA and proteasome expression body map. BMC Medical Genomics. 2018;**11**(1):36

[31] Pellegrino MA, Belvedere M, Pellegrino AG, Ferrone S. B peripheral lymphocytes express more HLA antigens than T peripheral lymphocytes. Transplantation. 1978;**25**(2):93-95

[32] Yamamoto F, Suzuki S, Mizutani A, Shigenari A, Ito S, Kametani Y, et al. Capturing differential allele-level expression and genotypes of all classical HLA loci and haplotypes by a new capture RNA-Seq method. Frontiers in Immunology. 2020;**11**:941

[33] Yarzabek B, Zaitouna AJ, Olson E, Silva GN, Geng J, Geretz A, et al. Variations in HLA-B cell surface expression, half-life and extracellular antigen receptivity. eLife. 2018;**7**:e34961

[34] MacDonald KS, Fowke KR, Kimani J, Dunand VA, Nagelkerke NJ, Ball TB, et al. Influence of HLA supertypes on susceptibility and resistance to human immunodeficiency virus type 1 infection. The Journal of Infectious Diseases. 2000;**181**(5):1581-1589

[35] Noreen HJ, Yu N, Setterholm M, Ohashi M, Baisch J, Endres R, et al. Validation of DNA-based HLA-A and HLA-B testing of volunteers for a bone marrow registry through parallel testing with serology. Tissue Antigens. 2001; **57**(3):221-229

[36] Leen G, Stein JE, Robinson J, Maldonado Torres H, Marsh SGE. The HLA diversity of the Anthony Nolan register. Hla. 2021;**97**(1):15-29

[37] Robinson J, Barker DJ, Georgiou X, Cooper MA, Flicek P, Marsh SGE. IPD-IMGT/HLA Database. Nucleic Acids Research. 2020;**48**(D1):D948-DD55

[38] HLA Nomenclature Database. Available from: https://hla.alleles.org/

[39] IPD-IMGT/HLA Database. Available from: https://www.ebi.ac.uk/ipd/imgt/hla/

[40] Nunes E, Heslop H, Fernandez-Vina M, Taves C, Wagenknecht DR, Eisenbrey AB, et al. Definitions of histocompatibility typing terms. Blood. 2011;**118**(23):e180-e183

[41] Nunes E, Heslop H, Fernandez-Vina M, Taves C, Wagenknecht DR, Eisenbrey AB, et al. Definitions of histocompatibility typing terms: Harmonization of Histocompatibility Typing Terms Working Group. Human Immunology. 2011;**72**(12):1214-1216

[42] Muller CA, Engler-Blum G, Gekeler V, Steiert I, Weiss E, Schmidt H. Genetic and serological heterogeneity of the supertypic HLA-B locus specificities Bw4 and Bw6. Immunogenetics. 1989; **30**(3):200-207

[43] Voorter CE, van der Vlies S, Kik M, van den Berg-Loonen EM. Unexpected Bw4 and Bw6 reactivity patterns in new alleles. Tissue Antigens. 2000;**56**(4): 363-370

[44] Bravo-Egana V, Sanders H, Chitnis N. New challenges, new opportunities: Next generation sequencing and its place in the advancement of HLA typing. Human Immunology. 2021;**82**(7):478-487

[45] Dunn PP. Human leucocyte antigen typing: Techniques and technology, a critical appraisal. International Journal of Immunogenetics. 2011;**38**(6):463-473

[46] Moyer AM, Gandhi MJ. Human leukocyte antigen (HLA) testing in pharmacogenomics. Methods in Molecular Biology. 2022;**2547**:21-45

[47] Allen ES, Yang B, Garrett J, Ball ED, Maiers M, Morris GP. Improved accuracy of clinical HLA genotyping by next-generation DNA sequencing affects unrelated donor search results for hematopoietic stem cell transplantation. Human Immunology. 2018;**79**(12): 848-854

[48] Lind C, Ferriola D, Mackiewicz K, Heron S, Rogers M, Slavich L, et al. Next-generation sequencing: The solution for high-resolution, unambiguous human leukocyte antigen typing. Human Immunology. 2010; **71**(10):1033-1042

[49] Marsh SGE. Nomenclature for factors of the HLA system, update July, August and September 2022. International Journal of Immunogenetics. 2022;**49**(6):385-416

[50] Marsh SGE, WHO Nomenclature Committee for Factors of the HLA System. Nomenclature for factors of the HLA system, update July, August and September 2022. Hla. 2022;**100**(6): 673-710

[51] Hu T, Chitnis N, Monos D, Dinh A. Next-generation sequencing technologies: An overview. Human Immunology. 2021;**82**(11):801-811

[52] National Marrow Donor Program (NMDP) Policies. Available from: https://bioinformatics.bethematchclinical.org/policies/

[53] Miller SA, Dykes DD, Polesky HF. A simple salting out procedure for extracting DNA from human nucleated cells. Nucleic Acids Research. 1988; **16**(3):1215

[54] Sanger F, Nicklen S, Coulson AR. DNA sequencing with chain-terminating inhibitors. Proceedings of the National Academy of Sciences of the United States of America. 1977;**74**(12): 5463-5467

[55] Tucker T, Marra M, Friedman JM. Massively parallel sequencing: The next big thing in genetic medicine. American Journal of Human Genetics. 2009;**85**(2): 142-154

[56] Cornaby C, Schmitz JL, Weimer ET. Next-generation sequencing and clinical histocompatibility testing. Human Immunology. 2021;**82**(11):829-837

[57] Brown NK, Merkens H, Rozemuller EH, Bell D, Bui TM, Kearns J. Reduced PCR-generated errors from a hybrid capture-based NGS assay for HLA typing. Human Immunology. 2021;**82**(4):296-301

[58] Logsdon GA, Vollger MR, Eichler EE. Long-read human genome sequencing and its applications. Nature Reviews. Genetics. 2020;**21**(10):597-614

[59] Loupy A, Lefaucheur C. Antibody-mediated rejection of solid-organ allografts. New England Journal of Medicine. 2018;**379**(12):1150-1160

[60] Tinckam KJ, Chandraker A. Mechanisms and role of HLA and non-HLA alloantibodies. Clinical Journal of the American Society of Nephrology. 2006;**1**(3):404-414

[61] Haas M, Loupy A, Lefaucheur C, Roufosse C, Glotz D, Seron D, et al. The Banff 2017 Kidney Meeting Report: Revised diagnostic criteria for chronic active T cell-mediated rejection, antibody-mediated rejection, and prospects for integrative endpoints for next-generation clinical trials. American Journal of Transplantation. 2018;**18**(2): 293-307

[62] Jaramillo A, Fernandez FG, Kuo EY, Trulock EP, Patterson GA, Mohanakumar T. Immune mechanisms in the pathogenesis of bronchiolitis obliterans syndrome after lung transplantation. Pediatric Transplantation. 2005;**9**(1):84-93

[63] Lefaucheur C, Viglietti D, Mangiola M, Loupy A, Zeevi A. From humoral theory to performant risk

stratification in kidney transplantation. Journal of Immunology Research. 2017; **2017**:5201098

[64] Loupy A, Hill GS, Jordan SC. The impact of donor-specific anti-HLA antibodies on late kidney allograft failure. Nature Reviews. Nephrology. 2012;**8**(6):348-357

[65] Maruyama T, Jaramillo A, Narayanan K, Higuchi T, Mohanakumar T. Induction of obliterative airway disease by anti-HLA class I antibodies. American Journal of Transplantation. 2005;**5**(9):2126-2134

[66] Robson KJ, Ooi JD, Holdsworth SR, Rossjohn J, Kitching AR. HLA and kidney disease: From associations to mechanisms. Nature Reviews. Nephrology. 2018;**14**(10):636-655

[67] Jin YP, Jindra PT, Gong KW, Lepin EJ, Reed EF. Anti-HLA class I antibodies activate endothelial cells and promote chronic rejection. Transplantation. 2005;**79**(3 Suppl): S19-S21

[68] Reinsmoen NL, Nelson K, Zeevi A. Anti-HLA antibody analysis and crossmatching in heart and lung transplantation. Transplant Immunology. 2004;**13**(1):63-71

[69] Terasaki PI, Ozawa M. Predicting kidney graft failure by HLA antibodies: A prospective trial. American Journal of Transplantation. 2004;**4**(3):438-443

[70] Valenzuela NM, Reed EF. Antibody-mediated rejection across solid organ transplants: Manifestations, mechanisms, and therapies. The Journal of Clinical Investigation. 2017;**127**(7): 2492-2504

[71] Jaramillo A, Smith CR, Maruyama T, Zhang L, Patterson GA, Mohanakumar T. Anti-HLA class I antibody binding to airway epithelial cells induces production of fibrogenic growth factors and apoptotic cell death: A possible mechanism for bronchiolitis obliterans syndrome. Human Immunology. 2003;**64**(5):521-529

[72] Sawyer GJ, Dalchau R, Fabre JW. Indirect T cell allorecognition: A cyclosporin A resistant pathway for T cell help for antibody production to donor MHC antigens. Transplant Immunology. 1993;**1**(1):77-81

Chapter 5

Tissue-Specific Immunity for Transplantable Endocrine Glands in the Context of HLA Expression

Beyza Goncu and Ali Osman Gurol

Abstract

Understanding the dynamic of the immune system, it is becoming clear that the characteristics of the tissue can be important as immune cells to determine the initiation and progression of an immune response. Among the various responses, tissue-specific immunity can be characterized by determining the Major Histocompatibility Complex (MHC). Human major histocompatibility antigens are known as the "human leukocyte antigen" (HLA) system. HLA contains more than 200 genes and has essential activities in immunology, diseases, and transplantation with gene regions of diverse functions. One of the significant roles in transplantation is donor and recipient selection. In allorecognition, once the recipient antigen-presenting cells (APC) recognize the donor tissue, this leads to activation and migration of the immune cells, which can promote rejection or tolerance. In solid organ transplantation, cultured tissue cells were presumed as passenger-leukocytes free, ensuring mainly prolonged graft survival. However, the current literature paves the way for understanding HLA peptides and allorecognition dynamics to prevent rejection or provide a definition for the donor-recipient match. Based on the given information, this chapter summarizes the HLA expression dynamics and allorecognition status from a transplantation perspective for endocrine glands, including the Adrenal glands, Pancreas, Parathyroid, and Thyroid glands.

Keywords: HLA, endocrine, adrenal gland, pancreas, parathyroid, thyroid

1. Introduction

The transplant immune response points out the importance of the characteristic features of the related tissues, triggering capacity, and continuation of the immune response. In order to understand the specific immune response of a particular tissue, it is necessary to determine the immunogenicity of that tissue. Immunogenicity is the expression of how one or more molecule stimulates the immune system and/or its ability to elicit a response. The concept of tissue immunogenicity can be defined and categorized by the Major Histocompatibility (MHC) antigens.

MHC antigens are Human Leukocyte Antigens (HLA). Related genes are located on the short arm of chromosome six, close to the centromere part, which occupies a four-megabase pair that encodes more than 200 gene regions [1, 2]. Proteins found on the cell surface or in the cytoplasm are examined in three groups according to their distribution, structure, and function in tissues; Class I, II, and III molecules. Class I includes HLA-A, -B, -C, -E, -F, -G, -H antigens, and Class II contains HLA-DR, -DP, -DQ, -DO, -DM antigens. In addition, Class III encodes various components involved in the complement system [3, 4].

HLA class I molecules -A, -B, -C are classical HLAs expressed in many tissues and found in almost all nucleated cells. HLA Class I molecules -E,-F,-G are designated non-classical HLAs [5, 6]. These molecules' primary purpose involves presenting peptide antigens to T cells [7]. When this situation is evaluated in terms of transplantation, the presented peptide is referred to as an allo-peptide. Moreover, the APCs of the recipient, presents the donor's peptides to T cells after transplantation. APCs mainly present peptides of HLA Class I molecules in two ways; transporter antigen processing complex (TAP) dependent or TAP-independent. While foreign (or allo-) peptides are transported from the endoplasmic reticulum to the cell membrane. During this subcellular process, ATP is not always required, whether TAP-dependent or TAP-independent [8].

HLA Class II molecules; -DR, -DP, -DQ are known as classical antigens, and -DO, -DM is defined as non-classical antigens. These exogenous antigens are expressed in B and T lymphocytes, macrophages, dendritic, epithelial cells, and endothelium [1]. For antigen presentation, these molecules enable foreign antigens (or allo-peptides) that enter those cells by phagocytosis or endocytosis [9]. The classical antigens of the HLA Class II molecules have heteronanomeres structures in the endoplasmic reticulum. These molecules are responsible for introducing the endosomal peptides from APCs to the T-cell membrane. In transplantation, the exogenous proteins (allo-molecules) are taken up by APCs via endocytosis. These proteins are degraded by lysosomal proteases/hydrolases. Usually, endogenous peptides carried by Class II antigens in vesicles produced in the normal cell cycle process are degraded over time and then excreted from the cell membrane after lysosomal degradation. However, in the presence of the exogenous peptides, the endogenous peptide attaches to the Class II molecule. Then exogenous peptides must compete before this degradation and binds it for presentation [10, 11].

Epithelial cells in the gastrointestinal tract express Class II molecules and act as APCs with particular CD74 positivity. In this mechanism, a trimer is formed by the binding of CD74 to the heterodimer Class II structure (α and β) within the endoplasmic reticulum (this structure is also known as the invariant chain or Class II histocompatibility antigen gamma chain). "class II-associated invariant chain peptide" (CLIP) is prepared for presentation in a short fragment-linked state (CLIP-HLA-Class II structure) by splicing a portion of CD74 by various proteases in the endosomal/lysosomal system. With the cleavage of the exogenous peptide-CLIP, it binds to the Class II molecule and is presented from the cell membrane surface [9, 12–15]. In this mechanism, the non-classical Class II molecule HLA-DM keeps Class II molecules stable within the endosome, until an exogenous peptide binds. In the presence of an exogenous peptide (if that peptide has sufficient affinity), it induces the CLIP to dissociate from the Class II molecule. After binding to the exogenous molecule, it rapidly presented through the cell membrane. Especially in B lymphocytes and dendritic cells, -DM must interact with -DO (non-classical Class II molecules). Afterward, the acidity level enhances proteolysis and promotes efficient peptide loading [12–17].

Based on the given information about the antigen presentation process, transplantation of composite tissues/cells or endocrine glands antigenic determination and presentation process differs depending on the HLA load of the donor's tissue.

Therefore, this chapter assesses and summarizes the current literature on HLA expression status for transplantable endocrine glands. Throughout this chapter, the main focus is the tissue expression profile contained in databases between all organs and particular endocrine glands. Additionally, the HLA expression studies performed by many researchers and their relationship with transplantations are reviewed.

2. Adrenal gland transplantation

The primary role of the adrenal gland is defined by its functional layers; zona glomerulosa, zona reticular, and the zona fasciculata. Each layer has distinct features and regulates various roles such as the reabsorption of sodium and water, production of sex hormones, mammalian stress response, etc. [18, 19].

Adrenal insufficiency was described in 1855 by Thomas Addison [20]. Since the developmental strategies in transplantation gained therapeutical advancements involving replacement therapies with glucocorticoids, and stem cell developments, those features have been combined by clinicians and physicians.

Transcriptomic Analysis of the adrenal gland showed the differential distribution of specific genes. Other genes that share with other organs are also classified by databases including The Human Protein Atlas (HPA), and The Genotype-Tissue Expression (GTEx) platforms (**Figure 1**) [22–25]. The adrenal gland has 24 specific and 220 elevated genes compared to other organs in humans. The development of the adrenal gland also contains 69 genes expressed in adult and fetal tissues [19]. Immunologically none of the genes are related to the HLA classes. The adrenal gland cells showed mononuclear infiltrations and high HLA expression in mRNA and protein levels for normal and diseased tissues [26]. In 1997, Marx et al. showed 17-hydroxylase expression in the zone fasciculata, and it started to express HLA Class II antigens in the early stages of life [27]. Afterward, in 2014 Leite et al. reported the relationship between the aggressiveness of pediatric adrenocortical tumors and low expression of HLA Class II molecules including -DRA, -DPA1, and -DPB1 [28]. A recent preprint from Altieri et al. determined immune clusters of the adult normal and adrenocortical adenomas in the human adrenal gland. They showed HLA class I (-A, -B, and -C) decreased in adenomas compared to the normal adrenal gland [29].

Adrenal gland transplantation is mainly performed as autotransplantation and in literature, approximately 52 studies were reported between 1948 and 2019. Among them, only four studies were written in languages other than English including Russian [30], Portuguese [31], Italian [32], and German [33], and the full text are not available among the other two cases as well [34, 35]. However, numerous in vivo studies evaluate autotransplantation, allotransplantation, and xenotransplantation approaches. Most of them were performed on rats [36–43], second on sheep [44–49], and third on dogs [50] and monkeys [51]. Clinically, routine hormone replacement therapy provides challenges over transplantation, and limited studies reported concise cases of adrenal transplants [52–54]. In literature, provided adrenal gland transplantation studies often contain in vivo studies and limited clinical cases. Considering the details, human embryonic stem cells, somatic cells such as fibroblast, hair follicles, and adipose tissue-derived stem cells were tested for personalized treatment options. Additionally, decellularization from cadaveric donors and subsequent recellularization from animal donor studies were performed in vivo only. Partial tissue allotransplantation in humans from related-living-donor was tested clinically [55]. Microencapsulation strategy is another tool for allotransplantation still, there are

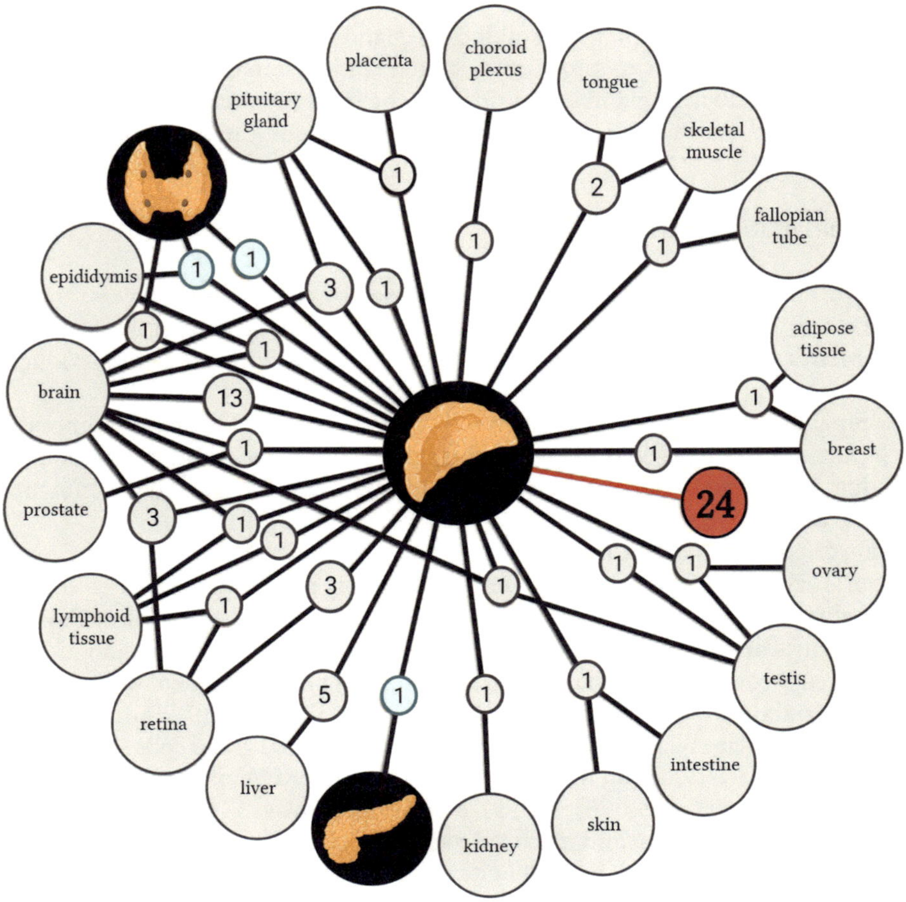

Figure 1.
Representative image of the particular shared genes and their relations between the adrenal gland and other organs. The image is reprinted from the human protein atlas database [21–23] and customized with BioRender.

no clinical studies for adrenal gland cells [56, 57]. Bornstein et al. recently achieved in vitro 3D models for bovine adrenal cells and provided viable organoid structures [58].

3. Pancreas transplantation

The pancreas is a unique organ that combines endocrine and exocrine functions, determining the variety of its pathology [59]. For a long time, it was believed that the ability of the pancreas to meet insulin requirements was minimal throughout life. In addition, evidence of plasticity of the pancreatic endocrine fraction has been reported, showing that the number of β-cells changes under certain physiological parameters. Besides, the endocrine function increases during changing metabolic demands such as obesity, or normal physiological growth. All basal cell types of the pancreas arise from a single pluripotent cell with a ductal phenotype [59, 60].

Transcriptomic Analysis of the pancreas provides a distribution of specific genes. The genes that share with the other organs are classified by databases involving HPA and GTEx platforms. The pancreas has 60 specific and 311 elevated genes compared to

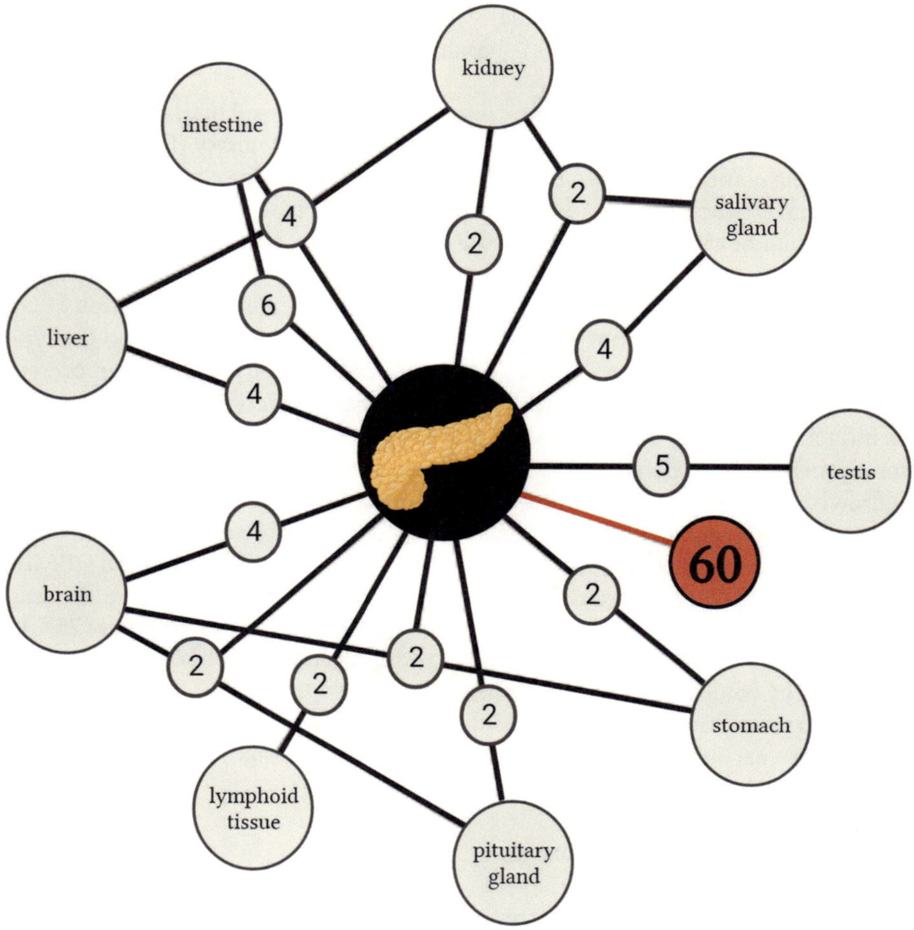

Figure 2.
Representative image of the particular shared genes and their relations between the pancreas and other organs. The image is reprinted from the human protein atlas database [21–23] and customized with BioRender.

other organs in humans (**Figure 2**) [23–25, 61], none of those reported genes involve HLA molecules. As a transplantable organ, the pancreas shows strong HLA expression profiles for HLA Class I and II. HLA-B and HLA-DRβ1 expressions were observed at high levels [25] and histologically, positivity could not be observed in either exocrine or endocrine cells [61]. Considering the possibility that HLA protein expression levels differ among pancreas tissues. HLA negative correlation between RNA profiling and histology does not guarantee a higher survival rate after transplantation.

The first total pancreas transplant in 1966 was performed by Kelly et al. [62]. After the first pancreas transplant; there was a gap due to poor transplant results, with the significant effects of poor organ preservation playing an important role [63]. The separation and transplantation of pancreatic islet cells became a reality many years later, after numerous experimental studies. In particular, over the past two decades, various efforts have made islet cell transplantation a viable therapy for many patients with type 1 diabetes (T1D) [63]. Current cell therapy gives patients with T1D the ability to have their insulin levels regulated by healthy endocrine functioning cells, rather than daily insulin injections. Clinical outcomes, including insulin independence, graft, and patient survival, have gradually improved and show results comparable to other organ transplants [64]. This is possibly due to the medical advancements in biotechnology [63, 65].

Between 2004 and 2013, studies on pancreas and islet transplantations continued throughout the world via the Collaborative Islet Transplant Registry (CITR). As suggested, transplantation should be performed according to the course of the disease and the clinical characteristics. Additionally, the type of diabetes and secondary etiologies should be considered [65]. The Leiden University Medical Center shared their 30 years of experience in 2015 and reported that they could observe an over 80% survival rate for 349 transplants [66]. Remarkably, factors such as the duration of cold ischemia of the tissue, the donor's age, the organ's procurement process, and the type of transplantation are essential in predicting the rejection risk [66]. Immunosuppressive treatments are offered to patients within two categories; induction and maintenance. In the induction part, the depletion of T cells is targeted, and in the maintenance part, a calcineurin inhibitor (CNI), and steroids are commonly used [65, 66]. There are two main drawbacks reported for islet transplantation compared to pancreas transplantation including the increase in the pancreas need and the risk of sensitization of the recipient [67]. Furthermore, Instant blood-mediated inflammatory reaction (IBMIR) is a well-known outcome regarding immune rejection and activation of complement after transplantation. Thus, it causes a severe loss of islet cells (approximately 25%) after vena cava infusion [68]. As is known, the Edmonton protocol provided promising outcomes by combining with immunosuppressive treatment [67, 69, 70]. Over the years, 15,000 procedures have already been performed so far [69, 71] with an outcome of one- to five-year survival rate with insulin independence reported [66, 69].

4. Parathyroid transplantation

The production of parathyroid hormone by the parathyroid glands is essential for controlling blood calcium levels. The human body needs precise calcium levels since even tiny variations can lead to issues with muscles and nerves [72]. If any or all of the parathyroid glands are underactive, individuals may have hypoparathyroidism. These four little glands exist in the body and they are situated near the thyroid gland in the neck region. The parathyroid gland keeps a balanced level of phosphorus, magnesium, and calcium in the blood. These glands do not produce enough parathyroid hormone if they are underactive (PTH). This primarily lowers the blood calcium level. All four parathyroid glands being damaged or removed is the most frequent cause. That may unintentionally occur during thyroid surgery. These glands are absent in some persons from birth. Alternatively, for unknown reasons, the glands do not function as well [72, 73].

Transcriptomic Analysis of the parathyroid gland presents a distribution of particular genes. In addition, different genes that share with other organs are classified by databases involving HPA and GTEx platforms. The parathyroid gland has 26 specific and 204 elevated genes compared to other organs in humans (**Figure 3**) [23–25].

Immunological properties of the parathyroid glands are evaluated in detail for HLA expression status for both tissue and isolated parathyroid cells [74–76]. In 1997, Tolloczko et al. compared the allotransplantation status of cultured parathyroid cells in the presence of IFNγ to decrease HLA Class I and II molecule expression [77]. With reduced HLA expression levels, parathyroid cells become more suitable for allotransplantation. However, based on their culture system, the survival rate after transplantation was reported to be 55% [78]. Flow cytometry outcome of these cells regarding

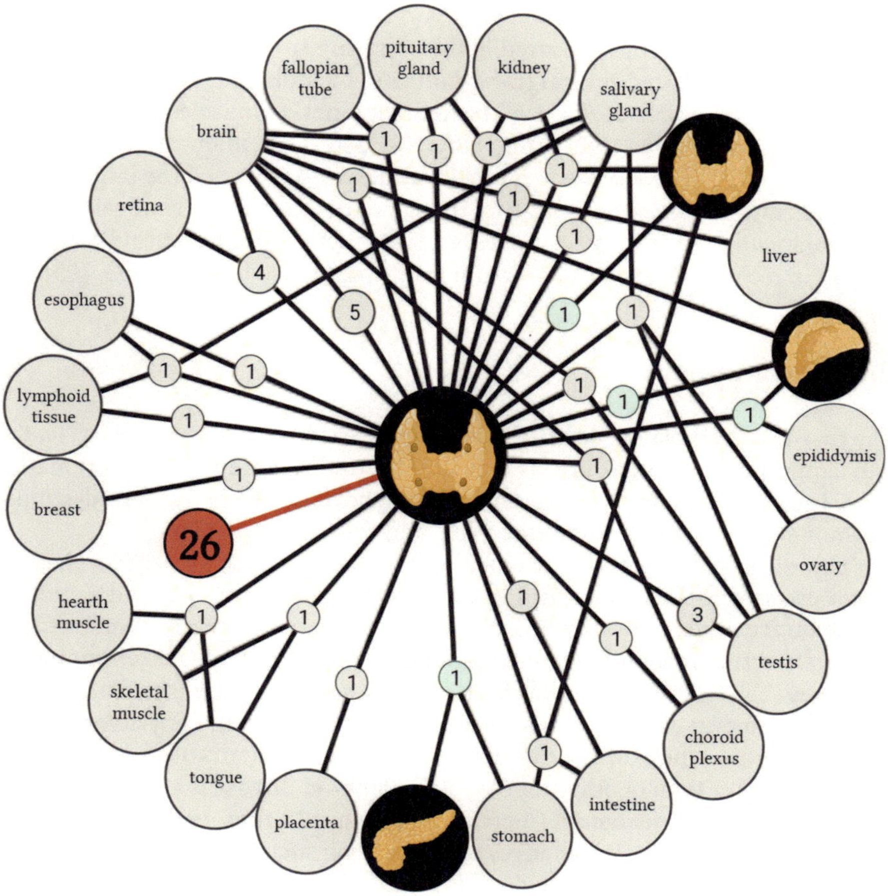

Figure 3.
Representative image of the particular shared genes and their relations between the parathyroid gland and other organs. The image is reprinted from the human protein atlas database [21, 23] and customized with BioRender.

HLA expression showed cells were negative for Class I molecules and positive for Class II [79]. However, an in vivo study showed that HLA Class I molecule expression in parathyroid cells causes more rapid rejection [80]. A recent study from Goncu et al. compared the HLA molecule expression with mRNA and protein levels via cultured cells [74, 75]. In their study, HLA-A molecule expression remained stable during 10 days of cultivation; also, HLA-B and -C expression could not detect at the protein level [75] despite the HPA histology staining data. HLA Class II molecules were also evaluated, and it was found that HLA-DP has higher mRNA expression levels, -DP, -DR, and -DQα1 protein expression levels showed a permanent expression among parathyroid tissues [74].

Parathyroid transplantation was first performed in 1911 by transplanting a tissue extract with thyroid tissue and the survival rate was reported to be up to 2 months [81]. The treatment of permanent hypoparathyroidism requires transplantation or hormone replacement therapy. From the first parathyroid transplantation, the 110 years to the present was demonstrated in various ways in literature, including survival rate, varying follow-up parameters, cell or tissue particle type of transplantation, and different transplant sites [82].

Autotransplantation is more frequently performed than allotransplantation [83]. There are 554 allotransplants reported in the literature from 1911 to 2021; among them, physicians evaluated several transplant sites including deltoid muscle [84–90], forearm muscle (brachialis) [78, 91–98], shoulder junction muscles (pectoralis major) [99], rectus muscle (rectus abdominis) [100, 101], sternocleidomastoid muscle [102], and omentum [103]. The Warsaw team in Poland, who performed the largest series of parathyroid transplants in the literature, reported that the intraperitoneal space acts as a natural incubator for parathyroid cells [104]. Therefore, parathyroid allotransplantation in recent studies uses omentum for its protective features [103, 105].

5. Thyroid transplantation

The thyroid gland plays a decisive role in homeostasis and development. The hypothalamic-pituitary-thyroid axis regulates thyroid function, and functional response is coordinated by thyrotropin-releasing hormone (TRH), thyroid-stimulating hormone (TSH), triiodothyronine (T3), and thyroxine (T4) [106].

The thyroid gland affects the metabolism of trace elements, and the level of trace elements also affects the metabolism and normal function of the thyroid gland. Changes in trace mineral levels will affect endocrine and other body systems causing thyroid dysfunction including hyperthyroidism, hypothyroidism, autoimmune thyroid diseases (Graves and Hashimoto's), thyroid cancer, and other systemic diseases [106]. Trace elements are essential for human survival and many physiological processes, including those of the thyroid gland, where concentrations of many trace elements are higher than in other tissues [106]. Thyroid disease is a common endocrine disorder and the incidence is increasing. As a result, thyroid disease is attracting attention and some regional challenges remain strong such as the Black Sea region and Turkiye being an endemic goiter region [107, 108]. Although several research studies report inconclusive outcomes about the relationship between trace minerals and thyroid diseases.

Transcriptomic Analysis of the thyroid gland presents a distribution of certain genes. Different genes that share with other organs are classified by databases involving HPA and GTEx platforms. The thyroid gland has ten specific and 171 elevated genes compared to other organs in humans (**Figure 4**) [22–25]. However, differential expression profiles in HLA molecules in thyroid glands show various changes by cancer progression. In 2007, one study reported that the frequency of Class II antigen expression in papillary thyroid cancer (PTC) samples, particularly the expression of HLA-DR/-DQ antigen, was found positive at 46.8 and 53.2%, respectively (n = 77). Clinicopathologically -DR and -DQ expressions were also present without nodal metastasis [109]. Notably, most differentiated thyroid tumors derived from thyroid epithelial cells are slow-growing cancers. Thyroid tumorigenesis is a complex process regulated by the activation of oncogenes, inactivation of tumor suppressors, and alterations in programmed cell death [110].

In 2017, Selieger et al. defined the role of the HLA Class II presentation pathway in tumors [111]. HLA expression (particularly Class II antigens) influences the tumor antigen (TA)-specific immune responses. Several studies and databases provided concordant information to the literature about the frequency of HLA molecules for tumor types [21, 76, 109, 111]. It is noteworthy that methodology and lab-to-lab variations such as used antibodies, the patients' population's characteristics, and the disease's molecular pathogenesis may reflect those differences [111].

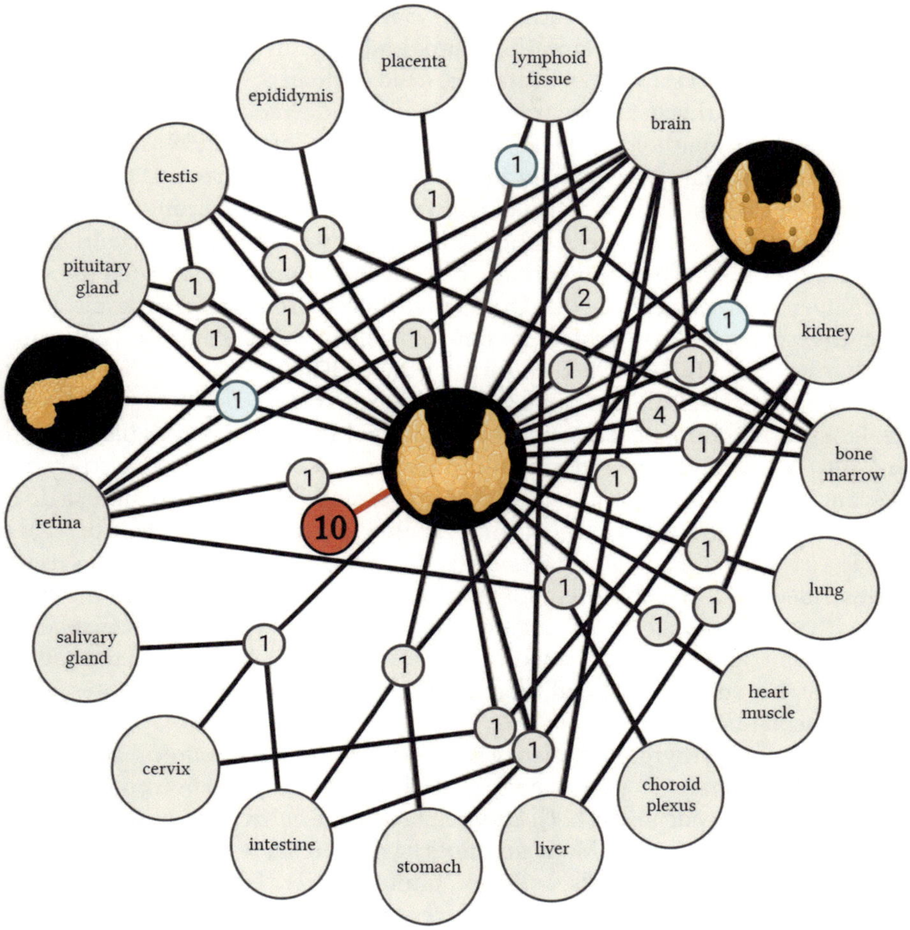

Figure 4.
Representative image of the particular shared genes and their relations between the thyroid gland and other organs. The image is reprinted from the human protein atlas database [21–23] and customized with BioRender.

The first record of thyroid transplantation in the literature belongs to the study of Albert Kocher in 1923 [112]. In patients with congenital thyroid insufficiency, administration of thyroid extract was used as a treatment of hypothyroidism, as a result, a certain level of success was achieved. A decrease in hypothyroidism symptoms was reported and 14% of the patients in this cohort (n = 204) showed otherwise [112]. Between 1946 and 1991 there were limited publications about thyroid transplantation studies written in languages other than English including Russian [113], French [114], and Japanese [115], and some of the studies did not have full text either. The first study that may be considered a starting point for thyroid transplantation was reported in 1978 by Perloff et al. The different sites were evaluated, and it was stated that the intrasplenic site was found to be the most unsuccessful transplantation area. Also, the thyroid was accepted as a "relatively privileged" tissue depending on the thyroid transplants that were considered successful. The originality of this study is that it is the first evaluation of thyroid and histocompatibility in terms of survival [116].

Approximately eight *in vivo* studies exist between 1981 and 2022 [117–123]. In these studies on thyroid transplantation; auto- and allotransplantation of irradiated thyroid tissue in rats with syngenic rats have been evaluated. At the 60-day follow-up, it was reported that thyroid tissue was functioning and irradiated thyroid auto-grafts would not show any pathological transformation [122]. Later, in 1988 Braun et al. reported an encapsulated transplantation of thyroid tissue *in vivo*, and the graft was placed beneath the kidney capsule. As an outcome thyroid encapsulated graft remained functional for 12 weeks [123]. Another study demonstrated a cultivation method for thyroid grafts in a hyperbaric oxygen condition and evaluated the survival, cellular infiltration, and HLA Class I molecule expressions. As a result, indicated cultivation systems provided immune cell depletion and decreased antigenicity of the thyroid graft [117]. At the beginning of the 2000s, Lee et al. provided donor-specific tolerance by administration of donor bone marrow cells via the portal venous route. Long-term follow-up was achieved and a combination of myeloablation therapy during 100-day survival was required for transplant success. That *in vivo* study has a significant role in stem cell and thyroid transplantation [119]. Furthermore, a different study reported the greater omentum feasibility as a transplant site for thyroid tissue via *in vivo* auto-transplantation model [120]. A more recent study from Wiseman et al. described a cell-pouch system for the thyroid tissue and determined the efficiency with viability and thyroid functionality after the subcutaneous transplantation model [121]. Intriguingly, the biomedical approach resulted in the fresh thyroid grafts showing a higher survival rate and functionality when compared to cryopreserved thyroid transplants [121].

The correction of thyroid deficiency remains a controversial clinical area despite the fact that more than 120 years having passed since the thyroid hormone was first used for therapeutic purposes [124]. Unfortunately, the cultivation of thyroid progenitor cells is a difficult task. Many attempts have been made to turn embryonic stem cells into functional thyroid cells, and here, limited success has been achieved mainly for mice [125]. Thyroid cell transplantation remains obscure for patients with congenital hypothyroidism and thyroid cancer. Besides, rejection occurs in the transplanted graft in patients with Hashimoto's thyroiditis or those with Graves' disease after the radioactive iodine or thyroidectomy process [125, 126].

6. Summary

Considering the antigen presentation and triggering an immune response after transplantation via HLA molecules, a common feature has been reported that 50% of endocrine tissues consist of non-immunogenic tissue [78, 94]. However, databases and existing studies report different expression outcomes for endocrine glands. When tissue and single-cell evaluations are mainly considered, two techniques come to the fore; histological evaluation and determination of RNA expression profile. **Figure 5** demonstrates HLA expression outcome in endocrine tissues, including the adrenal gland, pancreas, parathyroid, and thyroid for classical and non-classical HLA molecules. Among these transplantable endocrine glands, expression levels according to the organs are adrenal gland>thyroid gland>pancreas>parathyroid. Notably, the HLA-DRβ1 molecule expression is even higher than the other organs and other classes. Considering the expression of whole HLA Class I molecules expressions (both classical and non-classical molecules) were shown a decreasing pattern. On the contrary, this pattern lost its linearity and revealed the existence of different tissue-specific

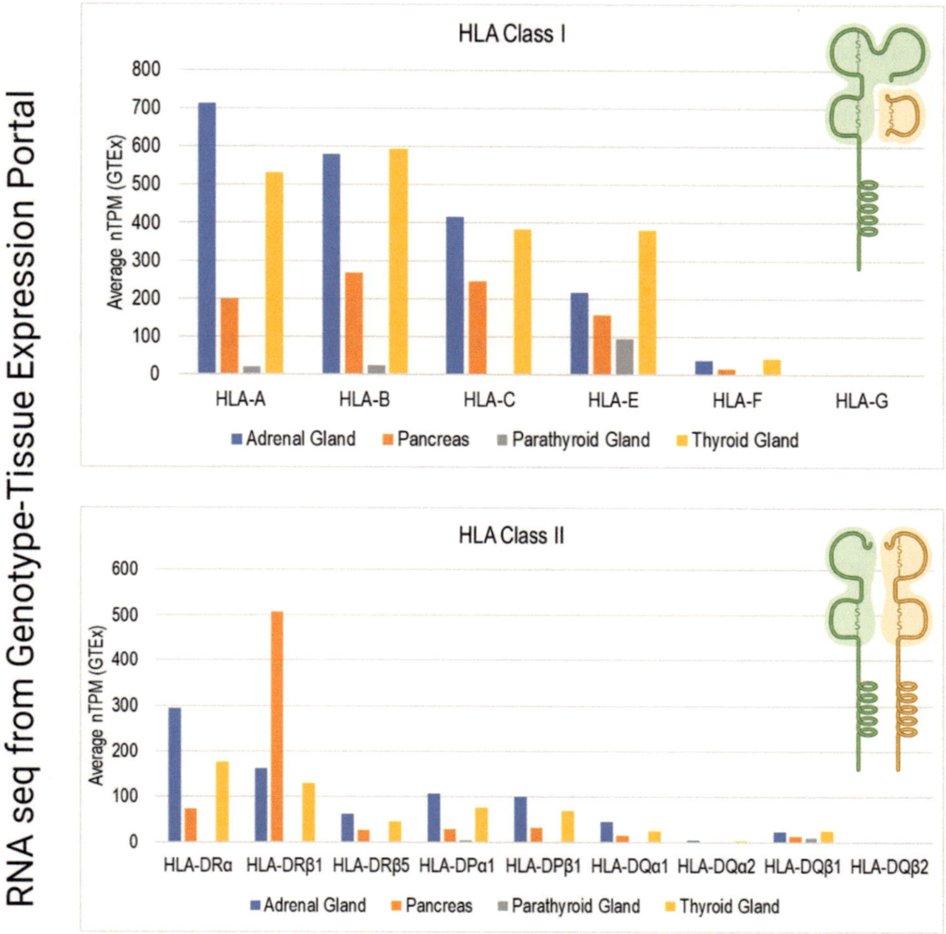

Figure 5.
HLA class I and II molecules expression in the particular endocrine organs involving the adrenal gland, pancreas, parathyroid, and thyroid gland, respectively. Expression levels refer to the number of transcripts per million (nTPM). RNA sequence expression data are retrieved from genotype-tissue expression portal (GTEx) [25]. Graphs were prepared in Microsoft excel and customized with BioRender.

profiles when Class II molecules were evaluated. Despite differences between expression profiles, the retrieval process causes a limitation in that the expression levels are observed in diseased tissue samples.

Difficulties in obtaining healthy tissues hinder researchers from studying with large cohorts. Hence, comparing the normal and diseased states of endocrine tissues becomes a challenge to render. Despite all this, studies were carried out with tissues from diseased individuals who already contribute to transplantation. As an example; the donor tissues used in the literature for the transplantation of parathyroid tissues belong to parathyroid hyperplasia tissues obtained from individuals who had chronic kidney failure. Another similar example; the pancreas transplantation, the endocrine cells rather than exocrine cells are preferred. Endocrine cell content is isolated and then transplanted. Although, deceased donors are preferred in the transplantation of thyroid and adrenal glands. The typical target in the transplantations of these endocrine tissues is the elimination of organ deprivation. Thus, the goal is to maintain metabolic responses more regularly instead of hormone replacement therapy.

This chapter outlines the topics mentioned above that may provide a solid ground to understand the HLA and endocrine gland interactions as well as some facilitating features for transplantation and crucial immunogenicity characteristics. Further, these comparisons elucidate that tissue-specific immunity has distinctive roles in transplantable endocrine glands.

Acknowledgements

This chapter is dedicated to all patients undergoing treatment for endocrine-related diseases who agreed to participate in the studies from many researchers highlighted in this section. This chapter was supported by the Research Fund Unit of Bezmialem Vakif University.

Conflict of interest

The authors declare no conflict of interest.

Author details

Beyza Goncu[1,2,3]* and Ali Osman Gurol[4,5]

1 Vocational School of Health Services, Department of Medical Services and Techniques, Bezmialem Vakif University, Istanbul, Turkey

2 Organ Transplant Center, Parathyroid Transplantation Unit, Bezmialem Vakif University Hospital, Istanbul, Turkey

3 Department of Immunology, Institute of Health Sciences, Istanbul University, Istanbul, Turkey

4 Department of Immunology, Aziz Sancar Institute of Experimental Medicine, Istanbul University, Istanbul, Turkey

5 Diabetes Application, and Research Center, Istanbul University, Istanbul, Turkey

*Address all correspondence to: bgoncu@bezmialem.edu.tr; bsgoncu@gmail.com

References

[1] Bray RA, Nickerson PW, Kerman RH, Gebel HM. Evolution of HLA antibody detection: Technology emulating biology. Immunologic Research. 2004;**29**(1-3):41-54

[2] Zabaneh D, Krapohl E, Simpson MA, Miller MB, Iacono WG, McGue M, et al. Fine mapping genetic associations between the HLA region and extremely high intelligence. Scientific Reports. 2017;**7**:41182

[3] Stavropoulos-Giokas C, Dinou A, Papassavas A. The role of HLA in cord blood transplantation. Bone Marrow Research. 2012;**2012**:485160

[4] Erlich H. HLA DNA typing: Past, present, and future. Tissue Antigens. 2012;**80**(1):1-11

[5] Crux NB, Elahi S. Human leukocyte antigen (HLA) and immune regulation: How do classical and non-classical HLA alleles modulate immune response to human immunodeficiency virus and hepatitis C virus infections? Frontiers in Immunology. 2017;**8**:832

[6] Marsh SG, Albert ED, Bodmer WF, Bontrop RE, Dupont B, Erlich HA, et al. Nomenclature for factors of the HLA system, 2010. Tissue Antigens. 2010;**75**(4):291-455

[7] Bontadini A. HLA techniques: Typing and antibody detection in the laboratory of immunogenetics. Methods. 2012;**56**(4):471-476

[8] Lawand M, Abramova A, Manceau V, Springer S, van Endert P. TAP-dependent and -independent peptide import into dendritic cell phagosomes. Journal of Immunology. 2016;**197**(9):3454-3463

[9] ten Broeke T, Wubbolts R, Stoorvogel W. MHC class II antigen presentation by dendritic cells regulated through endosomal sorting. Cold Spring Harbor Perspectives in Biology. 2013;**5**(12):a016873

[10] Hughes AL. Evolution of introns and exons of class II major histocompatibility complex genes of vertebrates. Immunogenetics. 2000;**51**(6):473-486

[11] Roche PA, Furuta K. The ins and outs of MHC class II-mediated antigen processing and presentation. Nature Reviews. Immunology. 2015;**15**(4):203-216

[12] van Lith M, McEwen-Smith RM, Benham AM. HLA-DP, HLA-DQ, and HLA-DR have different requirements for invariant chain and HLA-DM. The Journal of Biological Chemistry. 2010;**285**(52):40800-40808

[13] Baraliakos X, Baerlecken N, Witte T, Heldmann F, Braun J. High prevalence of anti-CD74 antibodies specific for the HLA class II-associated invariant chain peptide (CLIP) in patients with axial spondyloarthritis. Annals of the Rheumatic Diseases. 2014;**73**(6):1079-1082

[14] Adler LN, Jiang W, Bhamidipati K, Millican M, Macaubas C, Hung SC, et al. The other function: Class II-restricted antigen presentation by B cells. Frontiers in Immunology. 2017;**8**:319

[15] Turner JS, Marthi M, Benet ZL, Grigorova I. Transiently antigen-primed B cells return to naive-like state in absence of T-cell help. Nature Communications. 2017;**8**:15072

[16] Koch N, Zacharias M, Konig A, Temme S, Neumann J, Springer S.

Stoichiometry of HLA class II-invariant chain oligomers. PLoS One. 2011;**6**(2): e17257

[17] Denzin LK. Inhibition of HLA-DM mediated MHC class II peptide loading by HLA-DO promotes self tolerance. Frontiers in Immunology. 2013;**4**:465

[18] Stamou MI, Colling C, Dichtel LE. Adrenal aging and its effects on the stress response and immunosenescence. Maturitas. 2022;**168**:13-19

[19] Kempná P, Flück CE. Adrenal gland development and defects. Best Practice & Research Clinical Endocrinology & Metabolism. 2008;**22**(1):77-93

[20] Pearce JM. Thomas Addison (1793-1860). Journal of the Royal Society of Medicine. 2004;**97**(6):297-300

[21] Thul PJ, Akesson L, Wiking M, Mahdessian D, Geladaki A, Ait Blal H, et al. A subcellular map of the human proteome. Science. 2017;**356**(6340):1318-1330

[22] Bergman J, Botling J, Fagerberg L, Hallstrom BM, Djureinovic D, Uhlen M, et al. The human adrenal gland proteome defined by transcriptomics and antibody-based profiling. Endocrinology. 2017;**158**(2):239-251

[23] Uhlen M, Fagerberg L, Hallstrom BM, Lindskog C, Oksvold P, Mardinoglu A, et al. Proteomics. Tissue-based map of the human proteome. Science. 2015;**347**(6220):1260419

[24] Fagerberg L, Hallstrom BM, Oksvold P, Kampf C, Djureinovic D, Odeberg J, et al. Analysis of the human tissue-specific expression by genome-wide integration of transcriptomics and antibody-based proteomics. Molecular & Cellular Proteomics. 2014;**13**(2):397-406

[25] Consortium GT. The GTEx consortium atlas of genetic regulatory effects across human tissues. Science. 2020;**369**(6509):1318-1330

[26] Jackson R, McNicol AM, Farquharson M, Foulis AK. Class II MHC expression in normal adrenal cortex and cortical cells in autoimmune Addison's disease. The Journal of Pathology. 1988;**155**(2):113-120

[27] Marx C, Bornstein SR, Wolkersdörfer GW, Peter M, Sippell WG, Scherbaum WA. Relevance of major histocompatibility complex class II expression as a Hallmark for the cellular differentiation in the human adrenal cortex*. The Journal of Clinical Endocrinology & Metabolism. 1997;**82**(9):3136-3140

[28] Leite FA, Lira RC, Fedatto PF, Antonini SR, Martinelli CE, de Castro M, et al. Low expression of HLA-DRA, HLA-DPA1, and HLA-DPB1 is associated with poor prognosis in pediatric adrenocortical tumors (ACT). Pediatric Blood & Cancer. 2014;**61**(11):1940-1948

[29] Altieri B, Secener AK, Sai S, Fischer C, Sbiera S, Arampatzi P, et al. Cell atlas at single-nuclei resolution of the adult human adrenal gland and adrenocortical adenomas. bioRxiv. 2022. DOI: 10.1101/2022.08.27.505530 [Pre-print]

[30] Druzhinina KV. Modification of urinary corticoids in dogs following adrenal transplantation. Problemy Ėndokrinologii i Gormonoterapii. 1955;**1**(6):64-66

[31] Srougi M, Gittes RF, de Goes GM. Adrenal transplantation. AMB; Revista da Associação Médica Brasileira. 1979;**25**(10):353-358

[32] Cavallaro A, Vargiu L, Vecchi A. Experimental research on adrenal gland

transplantation. I. The autograft. Chirurgia e Patologia Sperimentale. 1970;**18**(3):125-143

[33] Lie TS, Nagel K. Adrenal gland transplantation—Yesterday and today. Die Medizinische Welt. 1973;**24**(22):937-939

[34] Bodey GP, Mioduszewska O, Gey GO. Studies on adrenal transplantation. I. Homologous and autogenous transplants. The American Journal of Anatomy. 1961;**108**:167-177

[35] Kohlenbrener RM, Sheridan JT, Steiner MM, Inouye T. Fetal adrenal transplantation attempt in juvenile Addison's disease. Pediatrics. 1963;**31**:936-945

[36] Butcher EO. Adrenal autotransplants with hepatic portal drainage in the rat. Endocrinology. 1948;**43**(1):30-35

[37] Engeland WC. Plasma adrenocorticotropin concentration is elevated in adult rats with neonatal adrenal transplants. Endocrinology. 1984;**114**(6):2160-2166

[38] Engeland WC. Pituitary-adrenal function after transplantation in rats: Dependence on age of the adrenal graft. The American Journal of Physiology. 1986;**250**(1 Pt 1):E87-E93

[39] Vendeira P, Magalhaes MM, Magalhaes MC. Autotransplantation of the adrenal cortex: A morphological and autoradiographic study. The Anatomical Record. 1992;**232**(2):262-272

[40] Trammer A, Kellnar S, Welsch U, Schwarz HP. Transplantation of fetal adrenal glands in syngeneic rat strains. European Journal of Pediatric Surgery. 1994;**4**(4):249-252

[41] Till H, Kellnar S, Bohm R, Schwarz HP, Barretton G, Joppich I. Fetal adrenal transplantation: Success of a laparoscopic technique in rats. Journal of Pediatric Surgery. 1997;**32**(10):1455-1457

[42] Till H, Stachel D, Muller-Hocker J, Joppich I. Cryopreservation and transplantation of fetal adrenal glands in adrenalectomized rats. European Journal of Pediatric Surgery. 1998;**8**(4):240-243

[43] Liu D, Qi S, Xu D, Zhu S, Chen H. A new vascularized adrenal transplantation model in the rat. Microsurgery. 2001;**21**(4):124-126

[44] Beaven DW, Espiner EA, Hart DS. The suppression of cortisol secretion by steroids, and response to corticotrophin, in sheep with adrenal transplants. The Journal of Physiology. 1964;**171**(2):216-230

[45] Goding JR, Wright RD. An improved method of preparing cervical adrenal transplants in the sheep. The Australian Journal of Experimental Biology and Medical Science. 1964;**42**:443-448

[46] Espiner EA, Hart DS, Beaven DW. Cortisol secretion during acute stress and response to dexamethasone in sheep with adrenal transplants. Endocrinology. 1972;**90**(6):1510-1514

[47] Lun S, Espiner EA, Hart DS. Adrenocortical metabolism of angiotensin in sheep with adrenal transplants. The American Journal of Physiology. 1978;**235**(5):E525-E531

[48] Lun S, Espiner EA, Hart DS. Effect of ACTH on the aldosterone response to potassium in sheep with adrenal transplants. The American Journal of Physiology. 1979;**237**(6):E500-E508

[49] Lun S, Espiner EA, Nicholls MG, Hart DS. Lack of direct effect of dopamine on aldosterone secretion in vivo. Endocrinology. 1983;**112**(1):60-63

[50] Dixit SP, Coppola ED. A new technique of adrenal transplantation by vascular anastomosis. The Journal of Surgical Research. 1969;**9**(1):42-48

[51] Gore AC, Saitoh Y, Terasawa E. Effects of adrenal medulla transplantation into the third ventricle on the onset of puberty in female rhesus monkeys. Experimental Neurology. 1996;**140**(2):172-183

[52] Dong D, Ji Z, Li H. Autologous adrenal transplantation for the treatment of refractory Cushing's disease. Urologia Internationalis. 2019;**103**(3):344-349

[53] Hardy JD. Autotransplantation of adrenal remnant to high in Cushing's disease. Preserving residual cortical activity while avoiding laparotomy. Journal of the American Medical Association. 1963;**185**:134-136

[54] Hardy JD, Moore DO, Langford HG. Cushing's disease today. Late follow-up of 17 adrenalectomy patients with emphasis on eight with adrenal autotransplants. Annals of Surgery. 1985;**201**(5):595-603

[55] Ruiz-Babot G, Hadjidemetriou I, King PJ, Guasti L. New directions for the treatment of adrenal insufficiency. Frontiers in Endocrinology (Lausanne). 2015;**6**:70

[56] Goncu B, Yucesan E. Microencapsulation for clinical applications and transplantation by using different alginates. In: Abu-Thabit APN, editor. Nano- and Micro-Encapsulation—Techniques and Applications. 1st ed. London, UK: IntechOpen Limited; 2020

[57] Balyura M, Gelfgat E, Ehrhart-Bornstein M, Ludwig B, Gendler Z, Barkai U, et al. Transplantation of bovine adrenocortical cells encapsulated in alginate. Proceedings of the National Academy of Sciences. 2015;**112**(8):2527-2532

[58] Bornstein S, Shapiro I, Malyukov M, Züllig R, Luca E, Gelfgat E, et al. Innovative multidimensional models in a high-throughput-format for different cell types of endocrine origin. Cell Death & Disease. 2022;**13**(7):648

[59] Jurij D, Viljem P, Marjan Slak R, Andraž S. Pancreas physiology. In: Andrada S, editor. Challenges in Pancreatic Pathology. Rijeka: IntechOpen; 2017. p. Ch. 2

[60] Alexandra EP, Yuliya SK, Larisa EG, Valeriy MB, Dmitriy AO, Iya AV, et al. Ontogeny of the human pancreas. In: Edward N, editor. Comparative Endocrinology of Animals. Rijeka: IntechOpen; 2019. p. Ch. 2

[61] Danielsson A, Ponten F, Fagerberg L, Hallstrom BM, Schwenk JM, Uhlen M, et al. The human pancreas proteome defined by transcriptomics and antibody-based profiling. PLoS One. 2014;**9**(12):e115421

[62] Squifflet JP, Gruessner RW, Sutherland DE. The history of pancreas transplantation: Past, present and future. Acta Chirurgica Belgica. 2008;**108**(3):367-378

[63] Wayne JH, Ahmer H, Henry P. Pancreas retrieval for whole organ and islet cell transplantation. In: Georgios T, editor. Organ Donation and Transplantation. Rijeka: IntechOpen; 2018. p. Ch. 9

[64] Barton FB, Rickels MR, Alejandro R, Hering BJ, Wease S, Naziruddin B, et al. Improvement in outcomes of clinical islet transplantation: 1999-2010. Diabetes Care. 2012;**35**(7):1436-1445

[65] Pedro V-A, Joana F-F, Maria JR. Pancreas transplantation. In: Georgios T, editor. Organ Donation and Transplantation. Rijeka: IntechOpen; 2018. p. Ch. 13

[66] Kopp WH, Verhagen MJ, Blok JJ, Huurman VA, de Fijter JW, de Koning EJ, et al. Thirty years of pancreas transplantation at Leiden University Medical Center: Long-term follow-up in a large Eurotransplant center. Transplantation. 2015;**99**(9):e145-e151

[67] Shapiro AM, Pokrywczynska M, Ricordi C. Clinical pancreatic islet transplantation. Nature Reviews. Endocrinology. 2017;**13**(5):268-277

[68] Eriksson O, Eich T, Sundin A, Tibell A, Tufveson G, Andersson H, et al. Positron emission tomography in clinical islet transplantation. American Journal of Transplantation. 2009;**9**(12):2816-2824

[69] Brennan DC, Kopetskie HA, Sayre PH, Alejandro R, Cagliero E, Shapiro AM, et al. Long-term follow-up of the Edmonton protocol of islet transplantation in the United States. American Journal of Transplantation. 2016;**16**(2):509-517

[70] Bottino R, Knoll MF, Knoll CA, Bertera S, Trucco MM. The future of islet transplantation is now. Frontiers in Medicine (Lausanne). 2018;**5**:202

[71] Castro-Gutierrez R, Michels AW, Russ HA. Beta cell replacement: Improving on the design. Current Opinion in Endocrinology, Diabetes, and Obesity. 2018;**25**(4):251-257

[72] Dawale K, Agrawal A. Parathyroid hormone secretion and related syndromes. Cureus. 2022;**14**(10):e30251

[73] Mannstadt M, Cianferotti L, Gafni RI, Giusti F, Kemp EH, Koch CA, et al. Hypoparathyroidism: Genetics and diagnosis. Journal of Bone and Mineral Research. 2022;**12**(37):2615-2629

[74] Goncu B, Yucesan E, Ersoy YE, Aysan ME, Ozten KN. HLA-DR, -DP, -DQ expression status of parathyroid tissue as a potential parathyroid donor selection criteria and review of literature. Human Immunology. 2022;**83**(2):113-118

[75] Goncu B, Yucesan E, Aysan E, Kandas NO. HLA class I expression changes in different types of cultured parathyroid cells. Experimental and Clinical Transplantation. Sep 2022;**20**(9):854-862. DOI: 10.6002/ect.2018.0388

[76] Natali PG, Bigotti A, Nicotra MR, Viora M, Manfredi D, Ferrone S. Distribution of human Class I (HLA-A,B,C) histocompatibility antigens in normal and malignant tissues of nonlymphoid origin. Cancer Research. 1984;**44**(10):4679-4687

[77] Tolloczko T, Wozniewicz B, Sawicki A, Gorski A. Allotransplantation of cultured human parathyroid cells: Present status and perspectives. Transplantation Proceedings. 1997;**29**(1-2):998-1000

[78] Barczynski M, Golkowski F, Nawrot I. Parathyroid transplantation in thyroid surgery. Gland Surgery. 2017;**6**(5):530-536

[79] Tsuji K, Fuchinoue S, Kai K, Kawase T, Kitajima K, Sawada T, et al. Culture of human parathyroid cells for transplantation. Transplantation Proceedings. 1999;**31**(7):2697

[80] Timm S, Hamelmann W, Otto C, Gassel AM, Etzel M, Ulrichs K, et al. Influence of donor MHC class I antigen expression on graft survival after rat parathyroid allotransplantation. Langenbeck's Archives of Surgery. 2001;**386**(6):430-433

[81] Brown WH. I. Parathyroid implantation in the treatment of Tetania Parathyreopriva. Annals of Surgery. 1911;**53**(3):305-317

[82] Ayşan ME, Göncü BS, Yücesan E. Paratiroit Nakli İmmünolojisi. Türkiye

Klinikleri, Kök Hücre ve Transplantasyon İmmünolojisi. 2019;**S1**(11):31-38. ISBN: 978-605-7597-98-4

[83] Miglietta F, Palmini G, Giusti F, Donati S, Aurilia C, Iantomasi T, et al. Hypoparathyroidism: State of the art on cell and tissue therapies. International Journal of Molecular Sciences. 2021;**22**(19):1-15

[84] Stone HB, Owings JC, Gey GO. Transplantation of living grafts of thyroid and parathyroid glands. Annals of Surgery. 1934;**100**(4):613-628

[85] Houghton BC, Klassen KP, Curtis GM. Studies in human parathyroid gland transplantation. Ohio State Medical Journal. 1939;**35**(5):505-508

[86] Flechner SM, Berber E, Askar M, Stephany B, Agarwal A, Milas M. Allotransplantation of cryopreserved parathyroid tissue for severe hypocalcemia in a renal transplant recipient. American Journal of Transplantation. 2010;**10**(9):2061-2065

[87] Aysan E, Altug B, Ercan C, Kesgin Toka C, Idiz UO, Muslumanoglu M. Parathyroid allotransplant with a new technique: A prospective clinical trial. Experimental and Clinical Transplantation. 2016;**14**(4):431-435

[88] Aysan E, Kilic U, Gok O, Altug B, Ercan C, Kesgin Toka C, et al. Parathyroid allotransplant for persistent hypocalcaemia: A new technique involving short-term culture. Experimental and Clinical Transplantation. 2016;**14**(2):238-241

[89] Yucesan E, Goncu B, Basoglu H, Ozten Kandas N, Ersoy YE, Akbas F, et al. Fresh tissue parathyroid allotransplantation with short-term immunosuppression: 1-year follow-up. Clinical Transplantation. 2017;**31**(11):1-4

[90] Aysan E, Yucesan E, Goncu B, Idiz UO. Fresh tissue parathyroid allotransplantation from a cadaveric donor without immunosuppression: A 3-year follow-up. The American Surgeon. 2020;**86**(4):e180-e1e2

[91] Feind CR, Weber CJ, Derenoncourt F, Williams GA, Hardy MA, Reemtsma K. Survival and allotransplantation of cultured human parathyroids. Transplantation Proceedings. 1979;**11**(1):1011-1016

[92] Torregrosa NM, Rodríguez JM, Llorente S, Balsalobre MD, Rios A, Jimeno L, et al. Definitive treatment for persistent hypoparathyroidism in a kidney transplant patient: Parathyroid allotransplantation. Thyroid. 2005;**15**(11):1299-1302

[93] Hasse C, Zielke A, Klock G, Barth P, Schlosser A, Zimmermann U, et al. First successful xenotransplantation of microencapsulated human parathyroid tissue in experimental hypoparathyroidism: Long-term function without immunosuppression. Journal of Microencapsulation. 1997;**14**(5):617-626

[94] Nawrot I, Wozniewicz B, Tolloczko T, Sawicki A, Gorski A, Chudzinski W, et al. Allotransplantation of cultured parathyroid progenitor cells without immunosuppression: Clinical results. Transplantation. 2007;**83**(6):734-740

[95] Cabané P, Gac P, Amat J, Pineda P, Rossi R, Caviedes R, et al. Allotransplant of microencapsulated parathyroid tissue in severe postsurgical hypoparathyroidism: A case report. Transplantation Proceedings. 2009;**41**(9):3879-3883

[96] Agha A, Scherer MN, Moser C, Karrasch T, Girlich C, Eder F, et al. Living-donor parathyroid allotransplantation for therapy-refractory postsurgical persistent hypoparathyroidism in a nontransplant

recipient—Three year results: A case report. BMC Surgery. 2016;**16**(1):51

[97] Agarwal A, Waghray A, Gupta S, Sharma R, Milas M. Cryopreservation of parathyroid tissue: An illustrated technique using the Cleveland clinic protocol. Journal of the American College of Surgeons. 2013;**216**(1):e1-e9

[98] Chapelle T, Meuris K, Roeyen G, De Greef K, Van Beeumen G, Bosmans JL, et al. Simultaneous kidney-parathyroid allotransplantation from a single donor after 20 years of tetany: A case report. Transplantation Proceedings. 2009;**41**(2):599-600

[99] Groth CG, Hammond WS, Iwatsuki S, Popovitzer M, Cascardo S, Halgrimson CG, et al. Survival of a homologous parathyroid implant in an immunosuppressed patient. Lancet. 1973;**1**(7812):1082-1085

[100] Bland WH, Chesson AL, Crow JB. Parathyroid gland transplantation; report of a case. North Carolina Medical Journal. 1950;**11**(9):501-504

[101] Garcia-Roca R, Garcia-Aroz S, Tzvetanov IG, Giulianotti PC, Campara M, Oberholzer J, et al. Simultaneous living donor kidney and parathyroid Allotransplantation: First case report and review of literature. Transplantation. 2016;**100**(6):1318-1321

[102] Giulianotti PC, D'Amico G, Tzvetanov I, Benedetti E. Living donor parathyroid allotransplantation with robotic transaxillary procurement in a kidney transplant recipient. Transplant International. 2014;**27**(6):e43-e45

[103] Goncu B, Salepcioglu Kaya H, Yucesan E, Ersoy YE, Akcakaya A. Graft survival effect of HLA-A allele matching parathyroid allotransplantation. Journal of Investigative Medicine. 2021;**69**(3):785-788

[104] Nawrot I, Wozniewicz B, Szmidt J, Sladowski D, Zajac K, Chudzinski W. Xenotransplantation of human cultured parathyroid progenitor cells into mouse peritoneum does not induce rejection reaction. Central-European Journal of Immunology. 2014;**39**(3):279-284

[105] Yucesan E, Basoglu H, Goncu B, Akbas F, Ersoy YE, Aysan E. Microencapsulated parathyroid allotransplantation in the omental tissue. Artificial Organs. 2019;**43**(10):1022-1027

[106] Zhou Q, Xue S, Zhang L, Chen G. Trace elements and the thyroid. Frontiers in Endocrinology (Lausanne). 2022;**13**:904889

[107] Erdogan G, Erdogan MF, Delange F, Sav H, Gullu S, Kamel N. Moderate to severe iodine deficiency in three endemic goitre areas from the Black Sea region and the capital of Turkey. European Journal of Epidemiology. 2000;**16**(12):1131-1134

[108] Kocakusak A. Did chernobyl accident contribute to the rise of thyroid cancer in Turkey? Acta Endocrinologica (Buchar). 2016;**12**(3):362-367

[109] Jo YS, Lee JC, Li S, Choi YS, Bai YS, Kim YJ, et al. Significance of the expression of major histocompatibility complex class II antigen, HLA-DR and -DQ, with recurrence of papillary thyroid cancer. International Journal of Cancer. 2008;**122**(4):785-790

[110] Yi HS, Chang JY, Kim KS, Shong M. Oncogenes, mitochondrial metabolism, and quality control in differentiated thyroid cancer. The Korean Journal of Internal Medicine. 2017;**32**(5):780-789

[111] Seliger B, Kloor M, Ferrone S. HLA class II antigen-processing pathway in tumors: Molecular defects and clinical relevance. Oncoimmunology. 2017;**6**(2):e1171447

[112] Kocher A. The treatment of hypothyroidism by thyroid transplantation. British Medical Journal. 1923;**2**(3274):560-561

[113] Khenkin VL. Homoplastic transplantation of the thyroid. Khirurgiia (Mosk). 1946;**5**:19-25

[114] Albeaux F, Alajouanine P, et al. Goiter by transplant. La Semaine des Hôpitaux. 1946;**22**(31 Suppl):1510

[115] Shimizu K, Kitamura Y, Nagahama M, Shoji T. A fundamental study of the thyroid transplantation for the patient with irreversible hypothyroidism (the first report: An autotransplantation of cryopreserved thyroid): Preliminary report. Nihon Geka Gakkai Zasshi. 1991;**92**(12):1728

[116] Perloff LJ, Utiger RD, Alavi A, Barker CF. Influence of histocompatibility and transplant site on survival. Annals of Surgery. 1978;**188**(2):186-196

[117] Hullett DA, Landry AS, Leonard DK, Sollinger HW. Enhancement of thyroid allograft survival following organ culture. Alteration of tissue immunogenicity. Transplantation. 1989;**47**(1):24-27

[118] Allen EM, Bartlett ST. The effect of methimazole, iodine and splenocytes on thyroid transplants in BB/Wor rats. Transplantation. 1999;**68**(1):25-30

[119] Lee S, Sugiura K, Nagahama T, Iwai H, Yasumizu R, Yamashita T, et al. New method for thyroid transplantation across major histocompatibility complex barriers using allogeneic bone marrow transplantation. Transplantation. 2001;**72**(6):1144-1149

[120] Gol'dberg ED, Popov OS, Udut VV, Stavrova LA, Galyan AN, Zhdanov VV, et al. A method for cellular stimulation of thyroid transplant for correction of thyroid system function. Bulletin of Experimental Biology and Medicine. 2005;**139**(4):529-531

[121] Wiseman SM, Memarnejadian A, Boyce GK, Nguyen A, Walker BA, Holmes DT, et al. Subcutaneous transplantation of human thyroid tissue into a pre-vascularized cell pouch device in a Mus musculus model: Evidence of viability and function for thyroid transplantation. PLoS One. 2022;**17**(1):e0262345

[122] Ahmed SA, Penhale WJ. Thyroid transplantation developing autoimmune thyroiditis following thymectomy and irradiation. Clinical and Experimental Immunology. 1981;**45**(3):480-486

[123] Braun K, Kauert C, Weber A, Hahn von Dorsche H, Hahn HJ. Syngeneic transplantation of thyroid tissue encapsulated in a cellulose sulphate membrane. Zeitschrift für Experimentelle Chirurgie, Transplantation, und Künstliche Organe. 1988;**21**(2):58-64

[124] Murray GR. The life-history of the first case of myxoedema treated by thyroid extract. British Medical Journal. 1920;**1**(3089):359-360

[125] Davies TF. Is thyroid transplantation on the distant horizon? Thyroid. 2013;**23**(2):139-141

[126] Lee MK, Bae YH. Cell transplantation for endocrine disorders. Advanced Drug Delivery Reviews. 2000;**42**(1-2):103-120

Chapter 6

Perspectives Concerning the Crucial Roles of MHC Molecules Corresponding to Individual Immunity against Emergent Viral Epidemics

Tirasak Pasharawipas

Abstract

Major histocompatibility complex (MHC) molecule is highly polymorphic. Each MHC allele has a limited capacity to bind its susceptible T cell epitopes to form an MHC-peptide complex (pMHC) and subsequently activates a compatible T cell clone. The cytotoxic T cell plays the role of eliminating virally infected cells to heal the infected patient. Overall, this article will discuss why some viral-infected patients can recover from the infected virus regardless of any treatment while others cannot. During the emergent viral epidemics, the new-coming medicines claimed to be able to treat the infected viruses. We should reconsider that some individuals, perhaps most, can recover from the infected virus with their own immunity. The fundamental comprehension of the critical role of the MHC molecule in individual immunity should be helpful to reconsider the strategy to manipulate the public health system including the notion to generate emergent viral vaccines.

Keywords: MHC, pMHC, MHC restriction, viral persistent infection, viral vaccine, viral epidemics

1. Introduction

Viral epidemics, such as severe respiratory syndrome (SARS), and COVID-19, have influenced our lives for a recent period of time. The infected diseases are caused by viruses classified as members of the Coronaviridae family [1]. In fact, all of the previous epidemic viruses such as the Dengue virus [2, 3], Japanese encephalitis virus [4], Influenza virus [5, 6] and many others [7–11] still keep causing global epidemics. Is it possible to find any effective medicines to cure the infected virus and eradicate them all? How do viruses still exist and cause epidemics? It is believed that adaptive immunity could act effectively by the role of cytotoxic T cell (Tc) to clear the viral infected cell. If so, why do the viruses keep causing problems for global public health? Perhaps, there are some things in the air that we should look for to reach our commitment to overcome these epidemics. We should reconsider our knowledge of where we are and how to move in the right direction.

Remarkably, this article will describe by differentiation of the terminology between viral invasion and viral infection. Viral invasion means our body is exposed to a particular virus which subsequently can induce our immunity with the immunogenic property of the virus. Viral infection means the virus is able to attach the cellular molecule with its receptor-binding domain and entry into the target cell. Thus, the host that does not have the compatible cellular molecules for the viral entry would not be infected but invaded. With the objective of the book and a limitation for the permitted space of the article, the details concerning the viral receptor molecules will not be discussed herein. With the questions mentioned above, this article will inform and discuss mainly the association of MHC molecules in relation to the individual's immunity to respond to the viral agents, which come out as survival or mortal.

2. Immuno-pathogenesis of viruses

Immune cells play a significant role in eliminating foreign substances that invaded a body. For clarification, immune cells respond to foreign agents as the non-selves substance. Immune cells cannot distinguish whether foreign substances are pathogenic or not. An example is allergens which act as foreign substances and cause allergy in some people although it is not a pathogen. The symptoms are based on the reaction of the immune cells by producing some kinds of immune substances, cytokines, to cause allergenic pathogenesis in those hypersensitive individuals [12, 13]. For most of the infected viruses, there is no evidence showing that the viruses directly cause pathogenesis. As same as allergens, the pathogenesis of the virally infected individuals is also caused by the roles of various cytokines that respond to the viral agents [14, 15]. However, a virus is different from an allergen in the way that it can regenerate in the compatible host but an allergen does not. The association between the viral load and the severity of the disease has been reported [16, 17]. Hence, viral replication enhances pathogenesis by inducing extremely high amounts of cytokines which are known as cytokine storms.

Reports show that most people worldwide have been exposed to some kinds of pathogenic viruses without self-notice. Various kinds of viral antibodies such as Dengue, Japanese encephalitis, yellow fever, etc. have been found in the major populations who have never been sick with the viruses [2–11]. World health organization (WHO) reported that 80% of the population who are positive for the Dengue serological tests are mild or asymptomatic and the mortality rate is lower than 1% [18]. For the Japanese encephalitis virus, approximately 0.4% of the seropositive individuals are severe [19]. This includes COVID-19 which approximately 70–85% of the SARS-CoV-2 positive detected are symptomless. About 10–15% are severe and require hospitalization while the mortality rate is 1–5% depending on the population and region [20, 21]. For those with severe symptoms, some can survive spontaneously without any specific treatment, while others cannot although the premium treatment has been manipulated. Aging and underlying diseases have been suggested to be the factors for the cause of severity and mortality [21–23]. However, there are reports of survival among aging and underlying patients. Thus, underlying diseases and aging might be a part of the reason but cannot be accounted as the key factor for the severity and mortality. In addition, new viral strains by the genomic mutation have also been suggested to be the reason to explain the different symptoms and severity [21, 22, 24]. However, the records show that the new viral strain, SARS CoV-2 (O-Micron) and many others still cause a similar ratio of asymptomatic, severity and mortality as the original SARS CoV-2 did. Similarly, immune evading of a virus is another mechanism to explain the cause of pathogenicity and severity including evading the previous immunity and vaccination. Questionably, if the virus evades immunity why it does not evade everybody? Why do not all vaccinated people get infected? These should be discussed.

3. Human immune response to viral antigen

After invasion into a body, the viral agent is captured by innate white blood cells such as macrophages and dendritic cells which play the role of the primary antigen-presenting cells (APCs). APCs digest and present the viral epitopes to induce adaptive immune cells which are cytotoxic T cell (Tc) and helper T cell (Th) clones in secondary lymphoid organs such as the lymph nodes and spleen [25–27]. APCs randomly cleave antigens into short peptides of 8–20 amino acid residues. The short peptide then combines with the MHC (major histocompatibility complex) molecule to form the MHC-peptide complex (pMHC) which plays a significant role to activate a specific T cell clone on its receptor so-called T cell receptor (TCR).

There are two classes of MHC molecules, class I and II. The MHC class I molecules can be expressed by any nucleated cells, while MHC class II molecules can be found only in the APCs. MHC is the key molecule to present the viral Ag on the cell surface of APC [25–27]. There are two pathways of antigen processing, the so-called endogenous [class I Ag processing] and exogenous [class II Ag processing] pathways. The endogenous pathway creates pMHC-I to activate a specific Tc cell clone [27–29]. The pMHC-I is a complex molecule of a short peptide of approximately 8–12 amino acids and MHC class I. pMHC-I is the crucial molecule to induce a specific naïve Tc cell clone to be an activated Tc. It then differentiates to effective Tc to attack the viral infected host cell. The exogenous pathway creates pMHC-II which comprises a larger epitope size, 12–20 amino acids, and MHC class II. The pMHC-II is a key to inducing the compatible Th cell clone [27, 29, 30]. The Th cell then plays the role to activate Tc to be an effective Tc.

In addition, the B lymphocyte is also one of the adaptive immune cells to play the main role extracellularly by synthesizing specific antibodies. Unlike the receptors of Tc and Th cells which recognize only short peptides, the B cell receptor (BCR) recognizes the conformational molecule of any substances which are not just proteins but also carbohydrates, lipids and nucleic acids. In other words, B lymphocyte does not need APCs to induce since it recognizes the specific epitope (B cell epitope) based on the native form of the antigen. The B cell epitope with approximately 5–20 amino acids of the native structure of the antigen can directly induce a B cell clone through the BCR [31]. After being induced, the activated B cell can synthesize IgM antibodies. Meanwhile, to differentiate to be plasma cells of each immunoglobulin class, which are IgG, IgA and IgE plasma cells, the B lymphocyte clone requires the cognate Th clone to promote [32]. During this period, the B and Th cells play the reciprocal role to support each other [31–33]. B cell which also expresses the MHC II molecule plays an antigen presentation role to the cognate Th which also sends some signals to promote B lymphocyte for differentiation to plasma cells to synthesize various classes of immunoglobulin. Without Th cells, B cells cannot produce other classes of immunoglobulin except IgM. More importantly, it cannot differentiate to be a memory B cell for prevention of the future invasion of the virus. IgM has a low affinity to bind the viral antigen and has limited action to combat the virus during extracellular existence. IgG is the highest capacity to bind the virus with its strong affinity. An affinity of IgA is the second best of IgG but plays a great role in mucosal organs which is the main route of many viral transmissions [34], including the SARS-CoV-2 virus. Accordingly, the Th cell plays the central role to maturate both B and Tc lymphocytes including the memory cells for long-term protection from the secondary viral infection [31–33].

4. The diversity of MHC molecules of human

MHC molecules are sets of molecules located on the cell membrane and are classified into 2 classes as mentioned. Each class has plural loci which are classified as the classical and non-classical loci. The MHC molecule of humans is HLA which stands for human leukocyte antigen (HLA) based on the fact that the MHC molecules were first found and studied in the white blood cell. Thus, HLA class I and II are synonyms for human MHC class I and II, respectively. The loci of classical HLA class I which have been known since 1950 are HLA-A, B and C while HLA-E, F and G were reported recently and have been called non-classical class I. In addition, the loci of classical HLA class II are HLA-DP, DQ and DR while those of the non-classical class II are HLA-DO and DM [26, 35]. The two classes of MHC molecules each have a different role in the adaptive cellular immune response. Each class of MHC genes comprises, at a minimum, three classical loci. MHC class I heterodimer is composed of alpha peptide and beta 2 microglobulins. Class I alpha peptide exhibits a high degree of polymorphism while beta 2 microglobulin does not. The two chains interact with one another non-covalently. As reported by the WHO Nomenclature Committee for Factors of the HLA System, the numbers of HLA-A, -B, and -C gene alleles and proteins are 7.644 (4.450), 9.097 (5.471), and 7.609 (4.218) $\times 10^3$, respectively [36]. HLA molecules are inherited co-dominantly from the parents. Thus, each locus of the MHC genome in an individual could be either heterozygous or homozygous. A heterozygous individual has two different gene alleles, while a homozygous has the same gene allele in the locus. Accordingly, the number of gene alleles of MHC class I in any individual are limited to 3–6 gene alleles. For example, the individual who has all three loci as homozygous would have only three gene alleles, while those who have all heterozygous loci would have six gene alleles. As the MHC gene alleles are highly polymorphic, the possibility of two individuals having the same set of gene alleles would not be less than one in a million (mostly, identical twin).

MHC class II molecules are also heterodimers of alpha and beta chains, coded by A and B genes, respectively. Both alpha and beta chains of MHC class II are highly polymorphism, HLA-DPB, -DQB and -DRB genes exhibit a much higher degree of polymorphism than the HLA-DPA, -DQA, and DRA genes. Thus far, the numbers of HLA-DPB1, -DQB1, and -DRB1 gene alleles and proteins are 2.221 (1.325), 2.330 (1.455), and 4.256 (2.828) $\times 10^3$, respectively while the numbers of alleles and proteins of DPA1, DQA1, and DRA are 491 (223), 508 (244), and 43 (5), respectively, in addition of a few gene alleles of DPB2, DQB2 [36]. Nevertheless, the combination of the alpha and beta peptides of HLA-II results in polymorphism more than HLA-I molecules.

5. The formation of pMHC molecule

The crystal structure of the MHC shows that each MHC molecule has a pocket to allow some of the amino acids of the peptide to fit in. As mentioned, individuals do not have more than six types for each class of MHC molecules. The studies found that each MHC variant has the ability to bind many different peptides [37, 38]. This makes MHC molecules have broad specificity to the T cell epitope presented by APCs. However, each MHC molecule can bind to only one peptide at once since there is only one cleft on an MHC molecule. To form pMHC, the MHC molecule requires only a few amino acids of T cell epitope peptide, the so-called anchor residue, for interaction. This allows each MHC allele to bind to many different peptides. Any processed peptides, which are derived from foreign substances by APCs, must contain the amino acids that can fit the MHC allele's cleft to form pMHC [39, 40]. Subsequently, the pMHC becomes the crucial molecule to induce a specific TCR of the T cell clone although there are many other molecules involving the interaction between the T cell clone and APC. TCR requires interaction with both the peptide residue and the MHC molecule [41]. This conforms to the development of T cell clones, in the thymus, which requires passing positive selection by having the ability to work with the self-MHC alleles of the individuals [42, 43]. Noticeably, reports showed that the interaction between each MHC allele and different peptides has a different affinity. Each MHC alleles have a limitation to bind to some of the peptides [44–46]. It is unlikely that all the epitopes of the foreign peptides are able to form pMHC with a single MHC allelic molecule (**Figure 1**). Accordingly, the MHC allelic molecules of each person have a limitation to forming pMHC with some of the peptides if those peptides do not contain the anchor residues that are compatible with the individuals' MHC alleles. There are studies showed that individuals who are MHC homozygous are more susceptible to pathogens than those who are heterozygous [47–49]. This explains that people who have fewer types of MHC alleles might have limitations to form pMHC molecules with some pathogen to induce naïve T cell clones. A good match of the peptide and the peptide-binding groove of the MHC molecule is a crucial factor in the induction of T-cell clones. Thus, the availability of the MHC allele and the antigen are the key MHC restriction factors for the induction of compatible T-cell clones. Besides the possibility of viral variants, a lack of available MHC alleles and antigens might explain why some individuals become infected and do not respond efficiently to gain seroprotection after viral vaccination [50, 51]. Therefore, the invasion of any particular antigen of a virus into different individuals does not guarantee the induction of the same level of immunity because of the limited varieties of MHC alleles in each person.

6. Immune evading versus chronic viral infection

Symptomatic treatment is a major strategy for treating most viral-infected patients until adaptive immunity is induced for the effective elimination of the viral agent. Usually, patients can recover from acute viral infections if their immune responses are produced within a specific period. On average, this requires one to two weeks. However, most viruses cause not only acute but also chronic infections. The definition of chronic viral infection is that the infected virus persists within a host for a longer period, usually longer than six months. During chronic infection, individuals might be asymptomatic but can transmit the virus to others [52, 53]. There are two significant types of chronic viral infections: latent and persistent. Latent infections

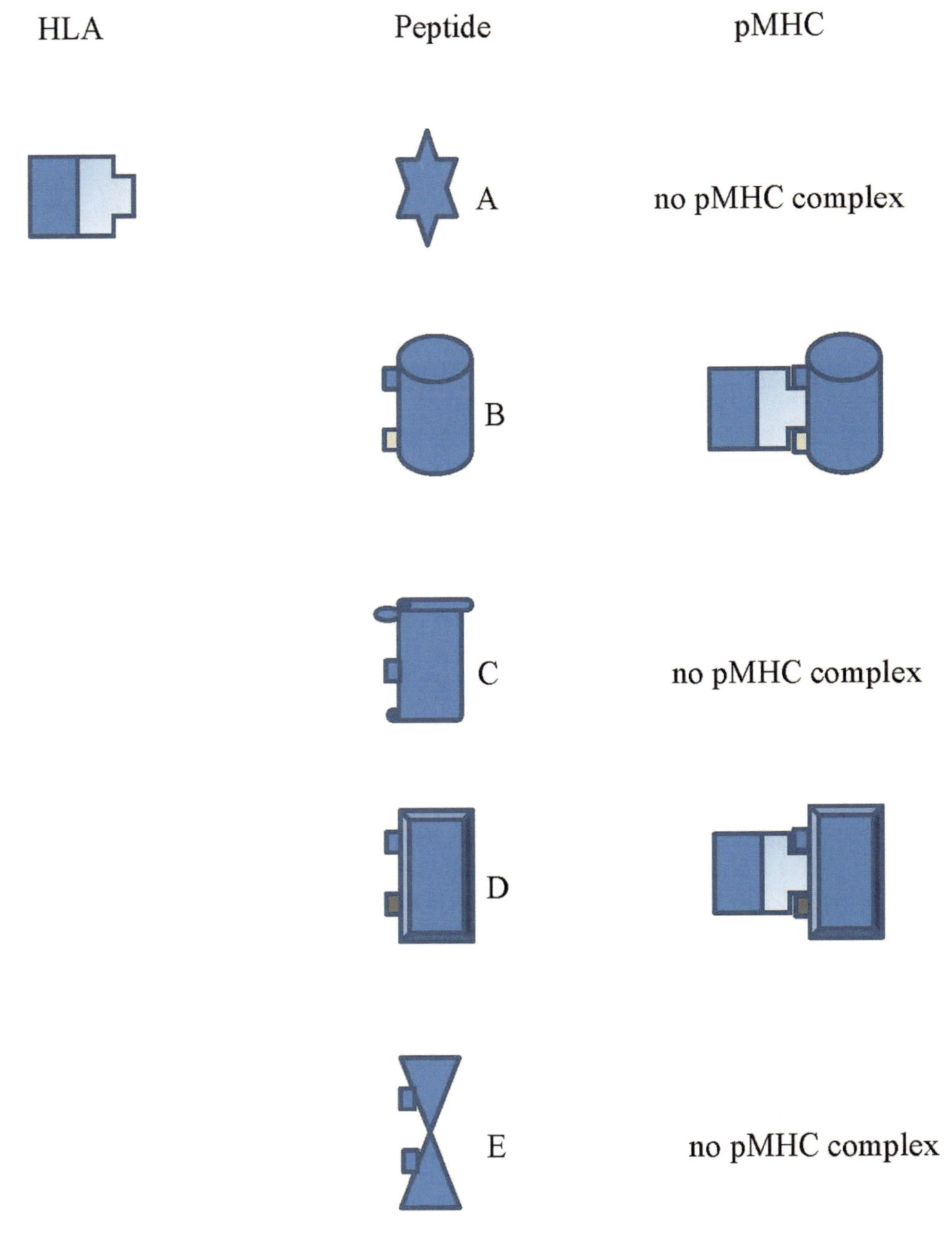

Figure 1.
In the illustration for the process to form the pMHC complex, the anchor residues of the epitope peptide must be compatible with the groove of the MHC allelic molecule to form pMHC. The HLA allele can form a pMHC complex to the peptides B and D which subsequently induce the TCR of each specific T cell clone but cannot do with the peptides A, C and E.

are caused by specific viruses that can evade the host's immunity after causing pathogenesis during the primary infection. Examples include members of Herpesviridae, such as herpes simplex and varicella-zoster [54, 55]. HSV produces latency-associated transcript (LAT) to inhibit cellular apoptosis to avoid the host's immunity. During this time, the virus can keep being dormant in its infected cell without any clinical symptoms to the host [56–58]. Besides the role of LAT, there were reports that herpes viruses also have a mechanism to interfere with the MHC class I expression of the target host cells [59]. Actually, this mechanism has been found in many other viruses as well. It explains that cytotoxic T cells [Tc] cannot recognize and eliminate the viral infected cell [60]. After acute HSV infection, the virus remains latently infected in all of the infected individuals who have not been treated properly during the primary acute infection. The viruses migrate to preferred cells or organs (such as nerve cells) where the host's immunity cannot respond. Acyclovir has been reported to be an effective antiviral drug to treat acute HSV infection to avoid chronically latent infection [61]. Accordingly, after acute HSV infection, the virus remains latently infected in the infected individuals who have not been treated or have not been treated properly during the primary acute infection. Eventually, patients usually experience latent infection [61, 62]. Thus, the mechanism of chronic infection of HSV should be accounted as immune evading which the mechanism seems to be explainable to all of the infected individuals and accounted as the viral factor.

The other chronic viral infection is viral persistence. The chronically persistent infection could be found in both DNA and RNA viruses. RNA viruses are much higher in their genomic mutation according to the low efficacy of their RNA polymerase to proofread their genomic replication [63, 64]. HCV, an RNA virus, is prone to cause a high prevalence of chronic infection. WHO reported that only about 30% (15–45%) of HCV-infected individuals have immune clearance within 6 months without any treatment. Of the rest of the HCV-infected individuals, approximately 70% (55–85%), developed a chronic infection [65]. However, many other RNA viruses have been reported to have a much lower prevalence of chronic infection e.g. Ebola [66], Influenza virus [67], Sars-Cov-2 [68], and other RNA viruses [69–72]. In addition, there were some reports showing that HCV mutation might not be the only cause of viral persistency [73, 74]. Accordingly, the genomic mutation of the virus might not be concluded as the only genuine cause of chronically persistent infection. HBV, a DNA virus, has been reported to cause chronic infection in approximately 10–15% of HBV-infected individuals [65]. Chronically infected HBV and HCV individuals tend to develop liver cirrhosis and hepatocellular carcinoma [75, 76]. Similar to Herpes viruses, HBV and HCV have also been reported to interfere with the MHC class I expression which was claimed to be the cause of their persistence [77].

Questionably, why do all the HBV and HCV-infected individuals not become chronically infected? As mentioned, approximately 85–90% of the HBV infected and 15–45% of the HCV infected can spontaneously clear the viral agent. As an example, the HBV and HCV persistently infected might not be explainable to be caused just by the virus factor as it does for the Herpes viruses. Like many other viruses including HSV, there are reports that HBV and HCV can interfere with MHC expression. In contrast to latent viral infection, chronically persistent viral infections vary by individual [69, 78–81]. Notably, the persistent viral agent continues to exist in the target organ of infected individuals without any major symptoms. Obviously, the infected host cannot clear the virus, and the infection becomes persistent. This raises the question of why chronic viral persistence cannot be found in all HBV and HCV-infected individuals as HSV does. Besides HBV and HCV, persistent viral infections can also be reported in many other viruses [82, 83]. Should one say that a virus factor such as MHC interference of viruses is the only factor to cause chronic viral infection? Should a host factor based on various genetics of individuals be the crucial key to explaining the cause of chronic viral persistent infections?

Accordingly, there are at least two major approaches to explain the cause of chronic viral infection. Firstly, it is a virus factor. The virus has a kind of mechanism to avoid host immunity. This kind of chronic infection should be the same in every individual as in the case of Herpesvirus which could be accounted for as the immune evading of the virus. The other approach is the different efficiency of the individual's immunity to fight against any particular infectious virus. In the latter case, some of the virally infected individuals can eventually clear the viral agents while others cannot and become persistently infected. This is expected to relate to the host factor, which is different and based on the individual's genetics. Thus, the polymorphic of MHC molecules is purposed to be the major factor herein.

7. The association of the MHC molecule and persistent viral infection

Viral persistent infection can also be found in a low-evolved immune animal. Insects, which lack adaptive immune response, many of which become viral carriers of Arboviruses, such as dengue hemorrhagic fever [84], Japanese encephalitis virus [85], West Nile virus [86], and others [69, 78–81]. The bee, which is an economic insect, can also be persistently infected with its pathologic viruses [87, 88]. Another example of viral persistent infection in low-evolved immune animals is penaeid shrimp which is a vital farming commodity in many countries. Thus, there are many reports concerning the viral epidemics in penaeid shrimp in addition to viral persistence.

Over two decades ago, the white spot virus and yellow head virus each caused pathogenesis and high mortality in infected shrimps when they emerged [89–91]. Notably, there is evidence of persistent viral infection in the shrimp, mortality declined sharply after several epidemic years once pond management systems such as water quality, feeding system, temperature, and rearing population size were optimized. In poorly managed farms, high mortality was reported [89, 91]. This could explain why shrimps perform a kind of unknown mechanism to tolerate infectious viruses. Alternatively, this might also be explained by the incidence of genomic mutation of the viruses which lowers their pathogenesis. The previous study, however, reported that the naïve shrimp showed acute infection and high mortality when challenged by a virus that was isolated from occluded, persistent-infected shrimp [89]. Accordingly, the viral genomic mutation should be excluded from the explanation for the cause of the persistent infection in shrimps. Insects and shrimps possess native immunity but not adaptive immunity. Naturally, innate immunity is not as sufficiently effective in clearing the virus and virus-infected cells as adaptive Tc.

In humans, there were reports showing the association of HLA variants and viral persistent infection. Bhaskaran et al. [92] reported that HLA-B*44 and DRB1*07 had a significant association with persistent HPV-16 infection (odds ratio, p-value = 26.3, 0.03 and 4.7, 0.01, respectively). HLA-B*27 and DRB1*12 were significantly associated with both HPV-16+ cervical cancer (CaCx) and persistent HPV-16 infection (23.8, 0.03; 52.9, 0.01; 9.8, 0.0009; and 13.8, 0.009; respectively). HLA-B*15 showed a negative association with HPV-16-positive CaCx (0.1, 0.01), whereas DRB1*04 exhibited protection to both HPV-16-positive CaCx and persistent HPV-16 infection (0.3, 0.0001 and 0.1, 0.0002, respectively). Besides, the associations of HLA variants and viral persistent infection of HBV [93–95] and HCV [96–98] have also been reported. These previous studies support the association between the HLA variants and the cause of viral persistent infection in some individuals which should be intended to evaluate the crucial role of MHC molecules concerning viral epidemics from the viral persistent carriers.

As mentioned, the host's immunity requires T cells, especially Tc and Th cells, for the effective clearance of the viral agent. MHC restriction is a key to activating T cell clones through their TCR. The antigenic epitope of the viral agent requires an MHC molecule to form the pMHC at the MHC groove and induce the compatible T lymphocyte clone. Each of the MHC alleles expresses different forms of the grooves [99, 100]. The limited capacity of the MHC groove to bind the viral short peptides results in lacking the appropriate pMHC to activate the significant T cell clone. In other words, the affinity differs between each MHC variant, and the distinct antigenic epitopes subsequently result in varying levels of the immune response. Hence, it assumes that persistently infected patients cannot clear the viral infected cell because they cannot produce an effective Tc cell for viral clearance. As is the case in infected shrimps, persistently infected patients do not have the adaptive immunity of the cellular-mediated immune response (CMIR). This could be due to the lack of an appropriate MHC-I allelic molecule to interact with the viral epitope, which results in the pMHC not forming to induce appropriate Tc lymphocyte clones. In addition, it is also interesting to raise the question of why the persistent virus can live in a host without any pathogenesis.

Besides activating, cytotoxic T cell also requires induction of Th for differentiation to an effective Tc for the efficient clearance of the infected cell, in addition, to producing memory cells for long-term activity. Thus, the compatible MHC-II is also an additional molecule to clear the virally infected cells. Individuals who lack either compatible MHC-I or II or both with the crucial viral epitopes cannot clear the viral infected cell efficiently. These infected patients might become severe cases unless the virus compromises to live in the host comfortably, as same as the infected shrimps that live in the optimal environment and come out asymptomatically [3, 100, 101]. The mechanism to compromise might be a process of adaptation of both virus and host, in addition to an environment. This subject has been discussed in many different aspects which require further study [102–104].

8. Perspective to increase the effectiveness of viral vaccines

As mentioned above, the MHC molecules play an important role in the immune response by forming the pMHC I and pMHC II complex molecule to activate Tc and Th cells, respectively, through the action of APC-containing cell clusters. Given that MHC class I and II molecules have many different alleles [36], while each of us has only 3–6 MHC alleles in each class, this is a limited binding to various epitopes. Therefore, some clones of T cell clones were not induced due to the lack of compatible pMHC [47–49]. This likely explains why a number of people do not have effective immunity after vaccination against HBV [9–14] and many other viruses [1–8]. It could be assumed that those vaccinated individuals lack the appropriate MHC allele to form the compatible pMHC complex with the vaccine epitopes. Thus, the required T cell clones could not be induced.

Th cells are important immune cells in helping to achieve the functioning of both Tc and B lymphocytes. Without a suitable Th cell cloning known as cognate Th cell, the activated B lymphocyte produces only IgM antibody, which has low stability and lower efficiency than IgG and IgA. Cognate Th cells can induce specific B cell clones to develop into plasma cells (antibody-secreting B cell) to produce other antibody classes (IgG, IgA, IgE) and memory B lymphocytes. For viral protection, antibodies are more important in preventing viral infection than Tc lymphocytes because the antibody can bind and inhibit the viral particles from binding to the viral receptor molecule, preventing the virus from entering the target cell and multiplying. This is known as antibody neutralization. Tc cells become active only after the target cell has been invaded by the virus since the recognition of the Tc cell clone through its TCR requires pMHC derived from virus-infected target cells before Tc cells can be manipulated. Therefore, the induction of memory B cells is an important goal of viral vaccines that the body's immune system must produce. Accordingly, viral vaccines should contain epitopes capable of binding to the MHC II alleles of individuals to form a pMHC-II as a capable molecule for inducing a suitable Th cell cognate clone. All these reasons suggest that any particular subunit viral vaccine may not be available to every community or the entire world. This is because viral vaccines should include an epitope appropriate for the type of MHC alleles in each person or population. Although the production of subunit viral vaccines is convenient with various technologies, especially with molecular genetic techniques, it raises problems and questions concerning the effectiveness of any particular vaccine for all global populations.

In the past, with much lesser knowledge concerning immunology, Dr. Edward Jenner and his associates had a way to initiate a successful battle against an epidemic of smallpox virus during the 18th century by the term of vaccination [105, 106]. Up till now, it seems we could not find any better outcome although a lot of medical scientific knowledge, including immunology, has been discovered. There are not sufficient kinds of viral vaccines to prevent the infection of many classic and emerging viruses. With a thought of safety reasons to prevent side effects, most of the scientists decided to work on subunit viral vaccines. This is different from the time when Edward Jenner used a whole particle of the Cowpox virus to vaccinate people to prevent the smallpox viral epidemic. On a different aspect, if the epitopes of any viral vaccines were cut down, an antigen-presenting cell would not be able to process induction to some particular helper T cell clones. Subsequently, the particular B cell clones cannot synthesize those particular antibodies to neutralize the infectious viral particles. Thus, could this be a reason that the subunit viral vaccines do not work in some people in the general public?

Some viral vaccines might show to be successful in the laboratory because the experimental animals were inbred to some specific strains. The viral vaccines that gave positive results at the experimental level do not mean it can work in human populations that contain various polymorphic MHC alleles. In addition, the viral vaccines that work in one population do not mean they always work in other populations. Perhaps, the subunit vaccine should be derived from all over the viral epitopes as if the wild-type virus. Of course, it is more appropriate to vaccinate people to gain viral protection on the receptor-binding domain (RBD). However, people who do not have the appropriate MHC alleles could not form the appropriate pMHC for the RBD which is the best target. Perhaps, their MHC's alleles might be able to form pMHC with other parts of the virus. This could allow them to synthesize memory B cells for some other parts of the virus. The memory B cell of other viral components might perform some actions to attack the infected virus. The combination of the antibody and natural killer (NK) cell by the antibody-dependent cell cytotoxicity (ADCC) mechanism could be helpful. This will be similar to Dr. Janner's manipulation and the manufacturing should not be too hard as we would produce the wild-type viral vaccines. This purpose should be considered for further study.

DOI: http://dx.doi.org/10.5772/intechopen.1000862

9. Conclusion

Some previous studies are convinced that chronically viral persistent infection is caused by viral factors such as viral mutation and some of the viral mechanisms. This article proposes that the cause of the chronic viral persistent infection should, also, associate with the existence of the MHC variants of the individual which is a host factor. Viral clearance requires the compatible MHC variants to induce Tc and Th for the effective eradication of the viral infected cell. There is no evidence that native immunity, such as natural killer cells and macrophages can clear the infectious virus efficiently by itself. The existence of a compatible MHC-I allele to the viral epitope is necessary for inducing the appropriate Tc clone to clear all the virally infected cells. The main fact is that the Tc clone [s] also requires compatible Th to support its efficiency. Thus, besides the MHC-I variants, the existence of the compatible MHC-II allele molecules is also another key to clearing the viral infected cell. The MHC molecules, as a key to preventing and restraining a viral epidemic, have been underestimated for their crucial role to eradicate the infected viral agent. Thus, this requires a better understanding of the distribution of MHC alleles in our global community. In addition, the Th cell is also a keystone for the regulation of immunoglobulin gene switching to synthesize more effective immunoglobulin which are IgG and IgA, not just IgM. More importantly, the Th cell is required for the B cell differentiation to be a memory B cell which makes viral prevention sustainable. Accordingly, the availability of the appropriate MHC class I manipulates the host to clear the viral infected cell effectively by the role of Tc. In addition, MHC class II is a key for the induction of the appropriate Th-cell clones to make adaptive immunity work properly in terms of prevention and clearance of the viral agent.

If the natural selection theory, proposed by Charles Darwin in 1858, influences creatures and human beings, the MHC gene could be one of the answers and become the target for us to overcome. Individuals have distinguished MHC allelic molecules which make each person respond to the viral agent differently. Any particular viral vaccines especially subunit viral vaccines do not seem to be available to prevent viral transmission. For viral treatment, it should have further studies to determine the MHC alleles for their association with the recovery of the viral infected patients. Using antiviral drugs, especially COVID-19 requires more careful evaluation. Those infected patients might be able to recover by their own immunity which relates to their MHC alleles to process Tc to clear the viral infected cell, not the medicine. If the medicine is truly effective to cure viral infected diseases why cannot it help many of the COVID-19 who do not have any other underlying diseases? Why is the mortality ratio of the COVID-19 epidemic still more or less the same although new variants are being found all the time? This needs an answer.

Acknowledgements

The author would like to appreciate Mr. Rohn Meyer for his kindness in helping proofread this article.

Author details

Tirasak Pasharawipas
Drug Discovery and Development Center, Office of Advanced Science and Technology, Thammasat University, Patumtani, Thailand

*Address all correspondence to: tirasak4124@yahoo.com; tirasak5@tu.ac.th

References

[1] Leao JC, Gusmao TPL, Zarzar AM, Leao Filho JC, Santos B, de Faria A, et al. Coronaviridae-old friends, new enemy! Oral Diseases. 2022;**28**(Suppl. 1):858-866. DOI: 10.1111/odi.13447. Epub 2020 Jun 26

[2] Luo S, Cui W, Li C, et al. Seroprevalence of dengue IgG antibodies in symptomatic and asymptomatic individuals three years after an outbreak in Zhejiang Province, China. BMC Infectious Diseases. 2018;**18**(1):92

[3] Ly S, Fortas C, Duong V, Benmarhnia T, et al. Asymptomatic dengue virus infections, Cambodia, 2012-2013. Emerging Infectious Diseases. 2019;**25**(7):1354-1362

[4] Nealon J, Taurel AF, Yoksan S, et al. Serological evidence of Japanese encephalitis virus circulation in Asian children from dengue-endemic countries. The Journal of Infectious Diseases. 2019;**219**(3):375-381

[5] Ip DK, Lau LL, Leung NH, et al. Viral shedding and transmission potential of asymptomatic and paucisymptomatic influenza virus infections in the community. Clinical Infectious Diseases. 2017;**64**(6):736-742

[6] Melchior TB, Perosa AH, Camargo CN, Granato C, Bellei N. Influenza virus prevalence in asymptomatic and symptomatic subjects during pandemic and postpandemic periods. American Journal of Infection Control. 2015;**43**(5):460-464

[7] Galanti M, Birger R, Ud-Dean M, et al. Rates of asymptomatic respiratory virus infection across age groups. Epidemiology and Infection. 2019;**147**:e176

[8] Attaran MS, Hosseini SM, Fakhari J, Sharifi Z. Serological and molecular characterization of hepatitis B virus in asymptomatic blood donors in Iran. Iranian Journal of Microbiology. 2018;**10**(1):59-64

[9] Glynn JR, Bower H, Johnson S, et al. Asymptomatic infection and unrecognised Ebola virus disease in Ebola-affected households in Sierra Leone: A cross-sectional study using a new non-invasive assay for antibodies to Ebola virus. The Lancet Infectious Diseases. 2017;**17**(6):645-653

[10] Stoszek SK, Engle RE, Abdel-Hamid M, Mikhail N, Abdel-Aziz F, Medhat A, et al. Hepatitis E antibody seroconversion without disease in highly endemic rural Egyptian communities. Transactions of the Royal Society of Tropical Medicine and Hygiene. 2006;**100**(2):89-94

[11] Sejvar JJ. West Nile Virus infection. Microbiology Spectrum. 2016;**4**(3). DOI: 10.1128/microbiolspec.EI10-0021-2016

[12] Rogala B, Bozek A, Gluck J, Jarzab J. Prevalence of IgE-mediated allergy and evaluation of Th1/Th2 cytokine profiles in patients with severe bronchial asthma. Postępy Dermatologii i Alergologii. 2015;**32**(4):274-280. DOI: 10.5114/pdia.2015.53323. Epub 2015 Aug 12

[13] Pelikan Z. Cytokine profiles in tears accompanying the secondary conjunctival responses induced by nasal allergy. Current Eye Research. 2014;**39**(2):120-132. DOI: 10.3109/02713683.2013.836542. Epub 2013 Oct 21

[14] Oldstone MB, Rosen H. Cytokine storm plays a direct role in the morbidity and mortality from influenza virus

infection and is chemically treatable with a single sphingosine-1-phosphate agonist molecule. Current Topics in Microbiology and Immunology. 2014;**378**:129-147. DOI: 10.1007/978-3-319-05879-5_6

[15] Yuan S, Jiang SC, Zhang ZW, Fu YF, Hu J, Li ZL. Quantification of cytokine storms during virus infections. Frontiers in Immunology. 2021;**12**:659419. DOI: 10.3389/fimmu.2021.659419

[16] Kwon JS, Kim JY, Kim MC, Park SY, Kim BN, Bae S, et al. Factors of severity in patients with COVID-19: Cytokine/chemokine concentrations, viral load, and antibody responses. The American Journal of Tropical Medicine and Hygiene. 2020;**103**(6):2412-2418. DOI: 10.4269/ajtmh.20-1110. Epub 2020 Oct 27

[17] Nakamura S, Iwanaga N, Hara S, Shimada S, Kashima Y, Hayasaka D, et al. Viral load and inflammatory cytokine dynamics associated with the prognosis of severe fever with thrombocytopenia syndrome virus infection: An autopsy case. Journal of Infection and Chemotherapy. 2019;**25**(6):480-484. DOI: 10.1016/j.jiac.2019.01.013. Epub 2019 Feb 26

[18] WHO. Dengue and Severe Dengue. 2022. Available from: https://www.who.int/news-room/fact-sheets/detail/dengue-and-severe-dengue

[19] WHO. Japenese Encephalitis. 2019. Available from: https://www.who.int/news-room/fact-sheets/detail/japanese-encephalitis

[20] WHO. 2022. Available from: https://covid19.who.int/?adgroupsurvey={adgroupsurvey}&gclid =Cj0KCQiAsdKbBhDHARIsANJ6-jdtyWU-UD_ZYkm0yVwxk-PEepbZ673LOJt4F6i6ceuMQbjgLTM7JhgaAkKwEALw_wcB

[21] Salzberger B, Buder F, Lampl B, Ehrenstein B, Hitzenbichler F, Holzmann T, et al. Epidemiology of SARS-CoV-2. Infection. 2021;**49**(2):233-239. DOI: 10.1007/s15010-020-01531-3. Epub 2020 Oct 8

[22] Williams N, Radia T, Harman K, Agrawal P, Cook J, Gupta A. COVID-19 severe acute respiratory syndrome coronavirus 2 [SARS-CoV-2] infection in children and adolescents: A systematic review of critically unwell children and the association with underlying comorbidities. European Journal of Pediatrics. 2021;**180**(3):689-697. DOI: 10.1007/s00431-020-03801-6. Epub 2020 Sep 10

[23] Bartleson JM, Radenkovic D, Covarrubias AJ, Furman D, Winer DA, Verdin E. SARS-CoV-2, COVID-19 and the ageing immune system. Nature Aging. 2021;**1**(9):769-782. DOI: 10.1038/s43587-021-00114-7. Epub 2021 Sep 14

[24] Meo SA, Meo AS, Al-Jassir FF, Klonoff DC. Omicron SARS-CoV-2 new variant: Global prevalence and biological and clinical characteristics. European Review for Medical and Pharmacological Sciences. 2021;**25**(24):8012-8018. DOI: 10.26355/eurrev_202112_27652

[25] Hume DA. Macrophages as APC and the dendritic cell myth. Journal of Immunology. 2008;**181**(9):5829-5835. DOI: 10.4049/jimmunol.181.9.5829

[26] Kelly A, Trowsdale J. Genetics of antigen processing and presentation. Immunogenetics. 2019;**71**(3):161-170. DOI: 10.1007/s00251-018-1082-2. Epub 2018 Sep 13

[27] Momburg F, Hengel H. Corking the bottleneck: The transporter associated with antigen processing as a target for immune subversion by viruses. Current Topics in Microbiology and Immunology. 2002;**269**:57-74. DOI: 10.1007/978-3-642-59421-2_4

[28] Lu X, Gibbs JS, Hickman HD, David A, Dolan BP, Jin Y, et al. Endogenous viral antigen processing generates peptide-specific MHC class I cell-surface clusters. Proceedings of the National Academy of Sciences of the United States of America. 2012;**109**(38):15407-15412. DOI: 10.1073/pnas.1208696109. Epub 2012 Sep 4

[29] Shastri N, Schwab S, Serwold T. Producing nature's gene-chips: The generation of peptides for display by MHC class I molecules. Annual Review of Immunology. 2002;**20**:463-493. DOI: 10.1146/annurev.immunol.20.100301.064819. Epub 2001 Oct 4

[30] Lázaro S, Gamarra D, Del Val M. Proteolytic enzymes involved in MHC class I antigen processing: A guerrilla army that partners with the proteasome. Molecular Immunology. 2015;**68**(2 Pt A):72-76. DOI: 10.1016/j.molimm.2015.04.014. Epub 2015 May 23

[31] Roche PA, Furuta K. The ins and outs of MHC class II-mediated antigen processing and presentation. Nature Reviews. Immunology. 2015;**15**(4):203-216. DOI: 10.1038/nri3818. Epub 2015 Feb 27

[32] Murin CD, Wilson IA, Ward AB. Antibody responses to viral infections: A structural perspective across three different enveloped viruses. Nature Microbiology. 2019;**4**(5):734-747. DOI: 10.1038/s41564-019-0392-y. Epub 2019 Mar 18

[33] Berzofsky JA. T-B reciprocity. An Ia-restricted epitope-specific circuit regulating T cell-B cell interaction and antibody specificity. Survey of Immunologic Research. 1983;**2**(3):223-229. DOI: 10.1007/BF02918417

[34] McHeyzer-Williams LJ, Malherbe LP, McHeyzer-Williams MG. Checkpoints in memory B-cell evolution. Immunological Reviews. 2006;**211**:255-268. DOI: 10.1111/j.0105-2896.2006.00397.x

[35] Rock KL, Reits E, Neefjes J. Present yourself! By MHC class I and MHC class II molecules. Trends in Immunology. 2016;**37**(11):724-737. DOI: 10.1016/j.it.2016.08.010. Epub 2016 Sep 7

[36] Marsh SGE. Nomenclature for factors of the HLA system, update July, August and September 2021. HLA. 2021;**98**(6):579-618. DOI: 10.1111/tan.14464

[37] Marzella DF, Parizi FM, van Tilborg D, Renaud N, Sybrandi D, Buzatu R, et al. PANDORA: A fast, anchor-restrained modelling protocol for peptide: MHC complexes. Frontiers in Immunology. 2022;**13**:878762. DOI: 10.3389/fimmu.2022.878762

[38] Perez MAS, Cuendet MA, Röhrig UF, Michielin O, Zoete V. Structural prediction of peptide-MHC binding modes. Methods in Molecular Biology. 2022;**2405**:245-282. DOI: 10.1007/978-1-0716-1855-4_13

[39] Rammensee HG. Chemistry of peptides associated with MHC class I and class II molecules. Current Opinion in Immunology. 1995;**7**(1):85-96. DOI: 10.1016/0952-7915[95]80033-6

[40] Nielsen M, Lundegaard C, Lund O. Prediction of MHC class II binding affinity using SMM-align, a novel stabilization matrix alignment method. BMC Bioinformatics. 2007;**8**:238. DOI: 10.1186/1471-2105-8-238

[41] Szeto C, Lobos CA, Nguyen AT, Gras S. TCR recognition of peptide-MHC-I: Rule makers and breakers. International Journal of Molecular Sciences. 2020;**22**(1):68. DOI: 10.3390/ijms22010068

[42] Lucas B, McCarthy NI, Baik S, Cosway E, James KD, Parnell SM, et al. Control of the thymic medulla and its influence on αβT-cell development. Immunological Reviews. 2016;**271**(1):23-37. DOI: 10.1111/imr.12406

[43] Tinsley KW, Hong C, Luckey MA, Park JY, Kim GY, Yoon HW, et al. Ikaros is required to survive positive selection and to maintain clonal diversity during T-cell development in the thymus. Blood. 2013;**122**(14):2358-2368. DOI: 10.1182/blood-2012-12-472076. Epub 2013 Aug 1

[44] Margulies DH, Corr M, Boyd LF, Khilko SN. MHC class I/peptide interactions: Binding specificity and kinetics. Journal of Molecular Recognition. 1993;**6**(2):59-69. DOI: 10.1002/jmr.300060204

[45] Sinigaglia F, Hammer J. Defining rules for the peptide-MHC class II interaction. Current Opinion in Immunology. 1994;**6**(1):52-56. DOI: 10.1016/0952-7915[94]90033-7

[46] Jensen KK, Andreatta M, Marcatili P, Buus S, Greenbaum JA, Yan Z, et al. Improved methods for predicting peptide binding affinity to MHC class II molecules. Immunology. 2018;**154**(3):394-406. DOI: 10.1111/imm.12889. Epub 2018 Feb 6

[47] Lipsitch M, Bergstrom CT, Antia R. Effect of human leukocyte antigen heterozygosity on infectious disease outcome: The need for allele-specific measures. BMC Medical Genetics. 2003;**4**:2. DOI: 10.1186/1471-2350-4-2. Epub 2003 Jan 24

[48] Arora J, Pierini F, McLaren PJ, Carrington M, Fellay J, Lenz TL. HLA heterozygote advantage against HIV-1 is driven by quantitative and qualitative differences in HLA allele-specific peptide presentation. Molecular Biology and Evolution. 2020;**37**(3):639-650. DOI: 10.1093/molbev/msz249

[49] Carrington M, Nelson GW, Martin MP, Kissner T, Vlahov D, Goedert JJ, et al. HLA and HIV-1: Heterozygote advantage and B*35-Cw*04 disadvantage. Science. 1999;**283**(5408): 1748-1752. DOI: 10.1126/science.283.5408.1748

[50] Faneye AO, Adeniji JA, Olusola BA, Motayo BO, Akintunde GB. Measles virus infection among vaccinated and unvaccinated children in Nigeria. Viral Immunology. 2015;**28**(6):304-308. DOI: 10.1089/vim.2014.0118. Epub 2015 Jun 23

[51] Smith LL, Milstone AM, Jibowu M, Luo CH, Morris CP, Mostafa HH, et al. Transmission of severe acute respiratory coronavirus virus 2 (SARS-CoV-2), delta variant, between two fully vaccinated healthcare personnel. Infection Control and Hospital Epidemiology. Dec 2022;**43**(12):1983-1985. DOI: 10.1017/ice.2021.469. Epub: 2021 Nov 8. PMID: 34743764; PMCID: PMC8632451

[52] Sharma SK, Saini N, Chwla Y. Hepatitis B virus: Inactive carriers. Virology Journal. 2005;**2**:82. DOI: 10.1186/1743-422X-2-82

[53] Furuya-Kanamori L, Yakob L. Heterogeneous and dynamic prevalence of asymptomatic influenza virus infections. Emerging Infectious Diseases. 2018;**24**(5):951. DOI: 10.3201/eid2405.180075

[54] Bello-Morales R, López-Guerrero JA. Extracellular vesicles in herpes viral spread and immune evasion. Frontiers in Microbiology. 2018;**9**:2572. DOI: 10.3389/fmicb.2018.02572

[55] Abendroth A, Kinchington PR, Slobedman B. Varicella zoster virus

immune evasion strategies. Current Topics in Microbiology and Immunology. 2010;**342**:155-171. DOI: 10.1007/82_2010_41

[56] Carpenter D, Hsiang C, Jiang X, Osorio N, BenMohamed L, Jones C, et al. The herpes simplex virus type 1 [HSV-1] latency-associated transcript [LAT] protects cells against cold-shock-induced apoptosis by maintaining phosphorylation of protein kinase B [AKT]. Journal of Neurovirology. 2015;**21**(5):568-575. DOI: 10.1007/s13365-015-0361-z. Epub 2015 Jun 13

[57] Jaggi U, Matundan HH, Tormanen K, Wang S, Yu J, Mott KR, et al. Expression of murine CD80 by herpes simplex Virus 1 in place of latency-associated transcript [LAT] can compensate for latency reactivation and anti-apoptotic functions of LAT. Journal of Virology. 2020;**94**(6):e01798-e01719. DOI: 10.1128/JVI.01798-19

[58] Imai Y, Apakupakul K, Krause PR, Halford WP, Margolis TP. Investigation of the mechanism by which herpes simplex virus type 1 LAT sequences modulate preferential establishment of latent infection in mouse trigeminal ganglia. Journal of Virology. 2009;**83**(16):7873-7882. DOI: 10.1128/JVI.00043-09. Epub 2009 Jun 3

[59] Piguet V. Receptor modulation in viral replication: HIV, HSV, HHV-8 and HPV: Same goal, different techniques to interfere with MHC-I antigen presentation. Current Topics in Microbiology and Immunology. 2005;**285**:199-217. DOI: 10.1007/3-540-26764-6_7

[60] Stone JD, Aggen DH, Chervin AS, Narayanan S, Schmitt TM, Greenberg PD, et al. Opposite effects of endogenous peptide-MHC class I on T cell activity in the presence and absence of CD8. Journal of Immunology. 2011;**186**(9):5193-5200. DOI: 10.4049/jimmunol.1003755. Epub 2011 Mar 30

[61] Kłysik K, Pietraszek A, Karewicz A, Nowakowska M. Acyclovir in the treatment of herpes viruses—A review. Current Medicinal Chemistry. 2020;**27**(24):4118-4137. DOI: 10.2174/0929867325666180309105519

[62] Menendez CM, Carr DJJ. Defining nervous system susceptibility during acute and latent herpes simplex virus-1 infection. Journal of Neuroimmunology. 2017;**308**:43-49. DOI: 10.1016/j.jneuroim.2017.02.020. Epub 2017 Mar 8

[63] Duffy S. Why are RNA virus mutation rates so damn high? PLoS Biology. 2018;**16**(8):e3000003. DOI: 10.1371/journal.pbio.3000003

[64] Malpica JM, Fraile A, Moreno I, Obies CI, Drake JW, García-Arenal F. The rate and character of spontaneous mutation in an RNA virus. Genetics. 2002;**162**(4):1505-1511. DOI: 10.1093/genetics/162.4.1505

[65] WHO. Hepatitis C. 2020. Available from: https://www.who.int/news-room/fact-sheets/detail/hepatitis-c

[66] Heeney JL. Ebola: Hidden reservoirs. Nature. 2015;**527**(7579):453-455. DOI: 10.1038/527453a

[67] Morens DM, Taubenberger JK, Fauci AS. The persistent legacy of the 1918 influenza virus. New England Journal of Medicine. 2009;**361**(3):225-229. DOI: 10.1056/NEJMp0904819. Epub 2009 Jun 29. Erratum in: New England Journal of Medicine. 2009;**361**(11):1123

[68] Kang H, Wang Y, Tong Z, Liu X. Retest positive for SARS-CoV-2 RNA of "recovered" patients with COVID-19: Persistence, sampling issues, or

re-infection? Journal of Medical Virology. 2020;**92**(11):2263-2265. DOI: 10.1002/jmv.26114. Epub 2020 Jun 9

[69] Doi T, Kwon HJ, Honda T, Sato H, Yoneda M, Kai C. Measles virus induces persistent infection by autoregulation of viral replication. Scientific Reports. 2016;**6**:37163. DOI: 10.1038/srep37163

[70] Hoarau JJ, Jaffar Bandjee MC, Krejbich Trotot P, Das T, Li-Pat-Yuen G, Dassa B, et al. Persistent chronic inflammation and infection by chikungunya arthritogenic alphavirus in spite of a robust host immune response. Journal of Immunology. 2010;**184**(10):5914-5927. DOI: 10.4049/jimmunol.0900255. Epub 2010 Apr 19

[71] Lin WH, Kouyos RD, Adams RJ, Grenfell BT, Griffin DE. Prolonged persistence of measles virus RNA is characteristic of primary infection dynamics. Proceedings of the National Academy of Sciences of the United States of America. 2012;**109**(37):14989-14994. DOI: 10.1073/pnas.1211138109. Epub 2012 Aug 7

[72] Meiring T, Prozesky L, Du Preez ER, Verwoerd DJ. The diagnosis and prevalence of persistent infection with bovine viral diarrhoea virus in south African feedlot cattle. The Onderstepoort Journal of Veterinary Research. 2011;**78**(1):323. DOI: 10.4102/ojvr.v78i1.323

[73] Fuller MJ, Shoukry NH, Gushima T, Bowen DG, Callendret B, Campbell KJ, et al. Selection-driven immune escape is not a significant factor in the failure of CD4 T cell responses in persistent hepatitis C virus infection. Hepatology. 2010;**51**(2):378-387. DOI: 10.1002/hep.23319

[74] Lapa D, Garbuglia AR, Capobianchi MR, Del Porto P. Hepatitis C virus genetic variability, human immune response, and genome polymorphisms: Which is the interplay? Cell. 2019;**8**(4):305. DOI: 10.3390/cells8040305

[75] Zhao SN, Liu LL, Lv ZP, Wang XH, Wang CH. Network analysis of HBV- and HCV-induced hepatocellular carcinoma based on Random Forest and Monte Carlo cross-validation. Molecular Medicine Reports. 2017;**16**(3):2411-2416. DOI: 10.3892/mmr.2017.6861. Epub 2017 Jun 27

[76] Urabe Y, Ochi H, Kato N, Kumar V, Takahashi A, Muroyama R, et al. A genome-wide association study of HCV-induced liver cirrhosis in the Japanese population identifies novel susceptibility loci at the MHC region. Journal of Hepatology. 2013;**58**(5):875-882. DOI: 10.1016/j.jhep.2012.12.024. Epub 2013 Jan 12

[77] Hewitt EW. The MHC class I antigen presentation pathway: Strategies for viral immune evasion. Immunology. 2003;**110**(2):163-169. DOI: 10.1046/j.1365-2567.2003.01738.x

[78] Boot HJ, Hahné S, Cremer J, Wong A, Boland G, van Loon AM. Persistent and transient hepatitis B virus [HBV] infections in children born to HBV-infected mothers despite active and passive vaccination. Journal of Viral Hepatitis. 2010;**17**(12):872-878. DOI: 10.1111/j.1365-2893.2009.01247.x

[79] Gokhale NS, Vazquez C, Horner SM. Hepatitis C virus. Strategies to evade antiviral responses. Future Virology. 2014;**9**(12):1061-1075. DOI: 10.2217/fvl.14.89

[80] Kanno T, Ishihara R, Hatama S, Uchida I. A long-term animal experiment indicating persistent infection of bovine coronavirus in cattle. The

Journal of Veterinary Medical Science. 2018;**80**(7):1134-1137. DOI: 10.1292/jvms.18-0050. Epub 2018 May 18

[81] Teunis PF, Sukhrie FH, Vennema H, Bogerman J, Beersma MF, Koopmans MP. Shedding of norovirus in symptomatic and asymptomatic infections. Epidemiology and Infection. 2015;**143**(8):1710-1717. DOI: 10.1017/S095026881400274X. Epub 2014 Oct 22

[82] McCarthy MK, Morrison TE. Persistent RNA virus infections: Do PAMPS drive chronic disease? Current Opinion in Virology. 2017;**23**:8-15. DOI: 10.1016/j.coviro.2017.01.003. Epub 2017 Feb 17

[83] Randall RE, Griffin DE. Within host RNA virus persistence: Mechanisms and consequences. Current Opinion in Virology. 2017;**23**:35-42. DOI: 10.1016/j.coviro.2017.03.001. Epub 2017 Mar 17

[84] Porter KR, Beckett CG, Kosasih H, Tan RI, Alisjahbana B, Rudiman PI, et al. Epidemiology of dengue and dengue hemorrhagic fever in a cohort of adults living in Bandung, West Java, Indonesia. American Journal of Tropical Medicine and Hygiene. 2005;**72**(1):60-66

[85] Thenmozhi V, Rajendran R, Ayanar K, Manavalan R, Tyagi BK. Long-term study of Japanese encephalitis virus infection in Anopheles subpictus in Cuddalore district, Tamil Nadu, South India. Tropical Medicine & International Health. 2006;**11**(3):288-293. DOI: 10.1111/j.1365-3156.2006.01567.x

[86] Nielsen CF, Reisen WK, Armijos MV, Maclachlan NJ, Scott TW. High subclinical West Nile virus incidence among nonvaccinated horses in northern California associated with low vector abundance and infection. The American Journal of Tropical Medicine and Hygiene. 2008;**78**(1):45-52

[87] Chen Y, Zhao Y, Hammond J, Hsu HT, Evans J, Feldlaufer M. Multiple virus infections in the honey bee and genome divergence of honey bee viruses. Journal of Invertebrate Pathology. 2004;**87**(2-3):84-93. DOI: 10.1016/j.jip.2004.07.005

[88] Locke B, Semberg E, Forsgren E, de Miranda JR. Persistence of subclinical deformed wing virus infections in honeybees following Varroa mite removal and a bee population turnover. PLoS One. 2017;**12**(7):e0180910. DOI: 10.1371/journal.pone.0180910

[89] Flegel TW. The shrimp response to viral pathogens. In: Proceedings of the Special Session on Sustainable Shrimp Aquaculture, World Aquaculture [190-214]. Orlando, Boca Raton, USA: World Aquaculture Society; 2001

[90] Sritunyalucksana K, Srisala J, McColl K, Nielsen L, Flegel TW. Comparison of PCR testing methods for white spot syndrome virus [WSSV] infections in penaeid shrimp. Aquaculture. 2006;**255**:95-104. DOI: 10.1016/j.aquaculture.2005.12.002

[91] Walker PJ, Mohan CV. Viral disease emergence in shrimp aquaculture: Origins, impact and the effectiveness of health management strategies. Reviews in Aquaculture. 2009;**1**(2):125-154. DOI: 10.1111/j.1753-5131.2009.01007.x. Epub 2009 May 15

[92] Bhaskaran M, Murali SV, Rajaram B, Krishnasamy S, Devasena CS, Pathak A, et al. Association of HLA-A, -B, DRB, and DQB alleles with persistent HPV-16 infection in women from Tamil Nadu, India. Viral Immunology. 2019;**32**(10):430-441. DOI: 10.1089/vim.2019.0094. Epub 2019 Dec 3

[93] Ramezani A, Aghakhani A, Kalantar E, Banifazl M, Eslamifar A,

Velayati AA. HLA-A *3303* and *3301 predispose patients to persistent hepatitis B infection. Journal of Gastrointestinal and Liver Diseases. 2009;**18**(1):117-118

[94] Pan N, Chen K, Qiu J, Sun H, Xu J, Miao F, et al. Human leukocyte antigen class I alleles and haplotypes associated with primary hepatocellular carcinoma in persistent HBV-infected patients. Human Immunology. 2013;**74**(6):758-763. DOI: 10.1016/j.humimm.2013.02.007. Epub 2013 Mar 1

[95] Zhu M, Dai J, Wang C, Wang Y, Qin N, Ma H, et al. Fine mapping the MHC region identified four independent variants modifying susceptibility to chronic hepatitis B in Han Chinese. Human Molecular Genetics. 2016;**25**(6):1225-1232. DOI: 10.1093/hmg/ddw003. Epub 2016 Jan 13

[96] Kondo Y, Ueno Y, Shimosegawa T. Dysfunction of immune systems and host genetic factors in hepatitis C virus infection with persistent normal ALT. Hepatitis Research and Treatment. 2011;**2011**:713216. DOI: 10.1155/2011/713216. Epub 2011 Jun 14

[97] McKiernan SM, Hagan R, Curry M, McDonald GS, Kelly A, Nolan N, et al. Distinct MHC class I and II alleles are associated with hepatitis C viral clearance, originating from a single source. Hepatology. 2004;**40**(1):108-114. DOI: 10.1002/hep.20261

[98] Mosaad YM, Farag RE, Arafa MM, Eletreby S, El-Alfy HA, Eldeek BS, et al. Association of human leucocyte antigen class I [HLA-A and HLA-B] with chronic hepatitis C virus infection in Egyptian patients. Scandinavian Journal of Immunology. 2010;**72**(6):548-553. DOI: 10.1111/j.1365-3083.2010.02468.x

[99] Fan S, Wang Y, Wang S, Wang X, Wu Y, Li Z, et al. Polymorphism and peptide-binding specificities of porcine major histocompatibility complex [MHC] class I molecules. Molecular Immunology. 2018;**93**:236-245. DOI: 10.1016/j.molimm.2017.06.024. Epub 2017 Jul 24

[100] Provenzano M, Panelli MC, Mocellin S, Bracci L, Sais G, Stroncek DF, et al. MHC-peptide specificity and T-cell epitope mapping: Where immunotherapy starts. Trends in Molecular Medicine. 2006;**12**(10):465-472. DOI: 10.1016/j.molmed.2006.08.008. Epub 2006 Sep 7

[101] Fares-Gusmao R, Rocha BC, Sippert E, Lanteri MC, Áñez G, Rios M. Differential pattern of soluble immune markers in asymptomatic Dengue, West Nile and Zika virus infections. Scientific Reports. 2019;**9**(1):17172. DOI: 10.1038/s41598-019-53645-w

[102] Flegel TW. Research progress on viral accommodation 2009 to 2019. Developmental and Comparative Immunology. 2020;**112**:103771. DOI: 10.1016/j.dci.2020.103771. Epub 2020 Jul 4. Erratum in: Developmental and Comparative Immunology. 2021;**118**:104016

[103] Pasharawipas T. Inducible viral receptor, a possible concept to induce viral protection in primitive immune animals. Virology Journal. 2011;**8**:326. DOI: 10.1186/1743-422X-8-326

[104] Zambon RA, Vakharia VN, Wu LP. RNAi is an antiviral immune response against a dsRNA virus in *Drosophila melanogaster*. Cellular Microbiology. 2006;**8**(5):880-889. DOI: 10.1111/j.1462-5822.2006.00688.x

[105] Barquet N, Domingo P. Smallpox: The triumph over the most terrible of the ministers of death. Annals of Internal

Medicine. 1997;**127**(8 Pt 1):635-642. DOI: 10.7326/0003-4819-127-8_part_1-199710150-00010. Erratum in: Annals of Internal Medicine. 1998;**128**(9):787

[106] Damaso CR. Revisiting Jenner's mysteries, the role of the Beaugency lymph in the evolutionary path of ancient smallpox vaccines. The Lancet Infectious Diseases. 2018;**18**(2):e55-e63. DOI: 10.1016/S1473-3099[17]30445-0. Epub 2017 Aug 18